57/Guns/888

HANDBOOK

FOR THE

ORDNANCE, M.L. 3-INCH MORTAR, MARK II

on Mounting, 3-inch Mortar, Mark I

LAND SERVICE

1937

By Command of the Army Council,

THE WAR OFFICE,
    31st August, 1937.

3

CONTENTS

LIST OF PLATES

LIST OF INSTRUCTIONAL WALL DIAGRAMS OF THE EQUIPMENT THAT ARE NOW AVAILABLE FOR ISSUE

Subject illustrated	*Diagram No.*
Ordnance, M.L. 3-inch Mortar, Mark II	E.262
Mounting, 3-inch Mortar, Mark I	E.263
Sight, 3-inch Mortar, Mark I	E.264
Ammunition—	
Bomb, M.L. 3-inch Mortar—H.E., smoke and practice	E.243
Bomb, M.L. 3-inch Mortar, 10-lb., Markings	E.478
Fuze, percussion, D.A., No. 138	E.242
Fuze, percussion, D.A., No. 150 and 150P, Mark I	E.471
Fuze, percussion, D.A., No. 152, Mark I	E.842

Demands for gun and carriage diagrams will be submitted to the C.O.O., Donnington and those for ammunition diagrams to the C.O.O., Bramley.

ORDNANCE, M.L. 3-INCH MORTAR, MARK II

ON

MOUNTING, 3-INCH MORTAR, MARK I

PREFACE

PART I

ORDNANCE

DEFINITION

The term " Ordnance " is applied to weapons designed for the propulsion of missiles by explosive force and includes guns, howitzers and mortars. Any one of such weapons may be referred to as a " piece " of ordnance.

Modern developments have rendered it somewhat difficult to draw any clearly defined line between the gun and howitzer. Generally speaking, for any given calibre, the gun is longer, has a more powerful propelling charge, higher muzzle velocity and a flatter trajectory than the howitzer or mortar. The mortar differs principally from the gun and howitzer in being lighter and simpler in construction ; it also employs a smaller propellant charge and lighter projectile. The chamber pressure and the range of the projectile are considerably less than with artillery equipments.

The mortar also differs from modern guns or howitzers in being breech or muzzle loading. With muzzle-loading mortars the projectile is rather smaller in diameter than the bore of the mortar with a consequent loss of propellant gases due to the windage.

The special characteristics of each type of weapon may be summarized thus :—

Gun	... ...	High velocity.	Long range.
Howitzer ...	...	Varying charges.	Steep angles of descent.
Mortar	... ...	Low velocity.	Short ranges.

Means of firing.—The breech end of the mortar barrel is fitted with a breech piece, which carries a striker stud projecting into the barrel. As the bomb slides down the bore on loading, a striker fitted over the cap of the primary cartridge in the base of the bomb is forced in, firing the cap, which in turn ignites the propellant charge. A disadvantage with this form of ignition is that there may be a missfire if the bomb does not slide freely down the bore during loading. It is, therefore, important that the bore of the mortar and the bands on the bomb should be free from dirt and rust.

PART II

THE MORTAR MOUNTING

The mounting required for mortars in the field must, of necessity, be of a light and portable nature, capable of easy transport and rapid assembly. For these reasons a mounting in the form of a bipod is employed as a support for the muzzle end of the mortar, the breech end being supported by a base plate. Facilities for increasing the angle of elevation and for a limited amount of traverse are incorporated in the mounting.

A sight is provided for laying the mortar and is designed to fulfil the dual purpose of a clinometer and reciprocating sight. A cross-level bubble enables the sight to be aligned in the vertical plane by cross-levelling, to compensate for difference in level of the legs and/or the effect of top traverse.

When a mortar is fired, the shock of discharge tends to cause the cradle of the bipod to move forward along the barrel of the mortar. To prevent the bipod being detached from the mortar and damaging the sight, the cradle is attached to the barrel by a powerful spring, and a stop band on the barrel ensures that the cradle is retained in the same position each time.

PART III

AMMUNITION

Ammunition for guns comprises four principal components :—

(a) The means of ignition.
(b) The propellant charge.
(c) The projectile.
(d) The fuze.

Although all equipments require to be provided as above, it does not follow that the ammunition used is of the same general type. In practice, differences arise between the ammunition used with B.L. and Q.F. guns and mortars respectively.

With B.L. guns the four parts of the complete round are, as a general rule, made up separately and are issued, stored and loaded independently of each other (except that shell and fuze are loaded together). The propellant charge is contained in a cloth bag and the means of ignition is a tube, either electric or percussion.

Q.F. guns, on the other hand, may have the ammunition made up to form one complete round to facilitate rapid loading, or it may be in two separate portions, one containing the means of ignition and the propellant charge, the other the projectile with fuze.

Mortars also have the ammunition made up in one complete round which can be loaded in one motion, as in the case of the Q.F. 3·7-inch mortar employed in tanks, or the cartridges may be secured to the vanes of the bomb, as in the case of the M.L. 3-inch mortar.

In the latter case, the means of ignition is the cap of a 95-grain cartridge of the sporting-gun type, which fits into a cylindrical-shaped chamber in the base of the bomb.

The propelling charge consists of two parts, namely, primary and secondary or augmenting. The primary charge is the 95-grain cartridge mentioned above, while the secondary charge consists of six celluloid cylinders or cambric bags filled with propellant explosive. The cylinders are placed between the vanes on the bomb, where they are retained in position by a retaining spring. The secondary cartridges augment the primary, the number used depending upon the range and the angle of descent required.

The mortar is equipped with the vaned type of bomb which may be filled with high explosive or smoke composition. Bombs, with a bursting charge of gunpowder, are also used for practice purposes. Vanes are necessary to ensure that the bomb will travel nose first, to give accuracy and to simplify fuze design. The bombs are detonated or exploded by means of a percussion type of fuze acting in conjunction with a suitable exploder or burster.

FUZES

Fuzes are used to initiate the explosion or detonation of the bursting charge of the bomb. They may be divided and subdivided into types depending upon the filling and mechanism.

From the filling point of view they may be divided into " igniferous " and " detonating " fuzes, the former designed to give a flash and the latter a detonating impulse.

Both types are provided with detonators. These consist of a small copper cup into which is pressed, with a fairly high density, about five grains of mercury fulminate or other detonating compositions. Detonators must be carefully filled and securely held in the fuze to prevent accidental ignition should the fuze or bomb be dropped or struck.

An *igniferous* fuze has a magazine of gunpowder and is employed when a flash only is required, as with the practice bomb with a gunpowder bursting charge.

The *detonating* fuze has the magazine filled with C.E. (composition exploding). As the detonation is initiated in the detonator, and continuity is essential between it and the magazine filling, there must of necessity be a channel filled with lightly stemmed C.E. between them.

This question of continuity rather complicates the internal structure of detonating fuzes, especially from the filling point of view.

Direct action fuzes are employed with the 3-inch mortar equipment. These fuzes may be of the igniferous or detonating types, the mechanism consisting of a needle supported on a thin metal disc, or a striker supported on a coiled spring, exposed to a direct blow ; its sensitivity depending on the strength of the disc or spring, respectively. Fuzes used with H.E. bombs are designed with shutters below the detonators.

EXPLOSIVES

An explosive is a substance which is capable of rapidly converting itself into gas with liberation of heat. To start this action a blow, friction, or the application of heat is generally necessary. Artillery explosives are usually solid substances.

Explosion is of two kinds :—

Simple explosion. Detonation.

The difference between these two lies in the rate at which the explosive is turned into gas. In a simple explosion the explosive substance is burnt layer by layer. The rate of burning is comparatively slow. Detonation is caused by the propagation of a disruptive wave through the explosive, with such great velocity—6,000 to 8,000 yards a second—that the effect is practically instantaneous and consequently severe. Simple explosion is influenced considerably by the amount of confinement.

Service explosives are divided into :—

1. Propellants (explosion rate only).

2. High explosives—
   (*a*) Initiating agents.
   (*b*) Intermediaries (or exploders).
   (*c*) Fillings (bursting charges).

3. Miscellaneous—
   Gunpowder, pyrotechnics, etc.

Propellants are used to impart velocity to the projectile, the propelling force being due to the pressure of the gas evolved by the burning. Explosives suitable for use as propellants are those in which the rate of burning is regular and, relatively speaking, slow. With such explosives, pressure may be controlled in the gun or mortar by means of the shape or size of the finished article.

High explosives, as far as artillery is concerned, consist of substances that may be readily detonated in a shell or bomb but which are yet sufficiently insensitive to withstand the shock of being fired from the gun or mortar.

Miscellaneous explosives include gunpowder and other explosives classed with it. They are used for general purposes in connection with explosive and pyrotechnic articles.

1. PROPELLANTS

Propellants used in the British service consist of gelatinized nitrocellulose, with or without nitroglycerine. Nitrocellulose is made by the action of nitric acid on cellulose, usually in the form of cotton waste or paper made from wood pulp. By the application of a solvent the fibrous nature of the nitrated cotton is destroyed and a jelly-like substance produced, which can be worked into any convenient shape and subsequently dried.

The solvents used are usually volatile and evaporate in the drying process, the nature of the solvent depending upon the degree of nitration of the explosive.

The propellants likely to be encountered are :—

Cordite.   N.C.T.   Ballistite.   N.C.(Y).

Cordite.—There are several types of cordite in the service, viz. cordite *Mark I*, cordite M.D., cordite M.C., cordite R.D.B., cordite S.C., cordite W. They are composed of nitroglycerine, nitrocellulose, mineral jelly or carbamite, the proportions in each case being as shown below :—

	Cordite Mk. I	Cordite R.D.B.	Cordite M.C.	Cordite M.D.	Cordite S.C.	Cordite W.
Nitroglycerine ...	58	42	30	30	41	29
Nitrocellulose (Insoluble) ...	37	—	65	65	—	65
Nitrocellulose (Soluble) ...	—	52	—	—	50	—
Mineral jelly ...	5	6	5	5	—	—
Carbamite ...	—	—	—	—	9	6

The insoluble nitrocellulose mentioned above is known in the service as guncotton. Insoluble nitrocellulose requires acetone as the solvent whereas, with soluble nitrocellulose, ether alcohol or other gelatinizing agents may be employed.

Properties

Cordite is practically smokeless and is also poisonous. It is much affected by the direct rays of the sun ; these cause it to deteriorate rapidly. Water or moisture do not affect it.

Cordite deteriorates progressively from the day it is made, the rate of deterioration depending on the storage temperature—the higher the storage temperature, the shorter the storage " life." This deterioration, however, does not appreciably affect the ballistics.

The temperature of the cordite charge at the moment of firing has a most important effect upon ballistics.

Cordite causes considerable erosion of the bore ; this effect is more pronounced with *Mark I*.

When fired to windward there may be a troublesome back flame when the breech is opened.

Cordite is comparatively difficult to ignite. For this reason all cordite cartridges for B.L. guns are fitted with gunpowder igniters.

Cordite is usually manufactured in the form of sticks or cords ; these may be solid or tubular. The cords may be sliced to form sliced cordite, or the cordite may be made in sheets and cut into squares, known as " square flake." The larger the diameter of the cord, i.e. its size, the slower it is consumed. Cordite is issued to the service in LOTS, the cordite in each lot having the same stability and ballistic coefficients.

N.C.T. is gelatinized nitrocellulose, with which a stabilizer is incorporated. It is made in the form of short, perforated sticks. Compared with cordite, N.C.T. has less power and requires a larger charge, is not so stable, consequently is not suitable for use in hot climates ; N.C.T. gives less flame but more smoke than cordite. A gunpowder igniter is used. N.C.T. should be kept dry because it absorbs moisture readily, which produces irregular ballistic results. N.C.T. has less erosive effect than cordite.

Ballistite consists of :—

Nitroglycerine ...	... ... 39·5
Nitrocellulose ...	... ... 60·5

Ballistite is smokeless and poisonous. It is affected by the direct rays of the sun ; these cause it to deteriorate rapidly.

Ballistite deteriorates progressively from the day it is made, the rate of deterioration depending on the storage temperature—the higher the storage temperature, the shorter the storage " life." This deterioration, however, does not appreciably affect the ballistics. It is rather difficult to ignite and requires a guncotton yarn priming between it and the cap in the primary cartridge.

N.C.(Y) is a trade explosive consisting of—

Nitrocellulose ...	 78
Barium nitrate ...	 12
Potassium nitrate	 2
Mineral jelly ...	 3
Woodmeal ...	 2
Starch ...	... ... 1·5
Camphor	... ... 1·5

It is in the form of small spherical grains, orange yellow in colour and is smokeless and poisonous.

The characteristics of N.C.(Y) are similar to ballistite regarding storage but, unlike ballistite, it contains small proportions of ingredients, e.g. mineral jelly and camphor, which have a stabilizing effect and so retard its deterioration.

2. HIGH EXPLOSIVES

The following are the high explosives used in artillery ammunition :—

(a) *Initiating agents* or detonators are used to initiate explosion or detonation and are usually employed in very small quantities, owing to the extreme sensitivity of the substances employed. Their use is confined to caps of primers and tubes, and detonators of fuzes.

Mercury fulminate is used alone in detonators of detonating fuzes or gaines. It is characterized by having great sensitiveness to impact and friction and is at the same time violent. To obtain full effect it must be confined. Any article containing mercury fulminate should be handled with great care. Water or oil will render it insensitive temporarily. It is mixed with other substances for use in non-detonating fuzes, primers, etc., the resultant product being known as detonating composition.

Lead azide has properties somewhat similar to those of mercury fulminate, but is less sensitive to percussion and may require a layer of detonating composition. It is much more stable in storage than fulminate, being unaffected by moisture or hot climates. The density is high and a smaller amount is required for a detonator than is the case with fulminate. Carbon dioxide in the presence of moisture may attack it.

(b) *Intermediaries* or exploders are employed to amplify the impulse given by the fuze.

C.E. is a pale yellow, crystalline substance and is used in fuzes and exploders for initiating the detonation of the main charge.

It is not affected by the exudation from amatol, and is more sensitive to shock and more responsive to initiation by fulminate or lead azide than either picric acid or T.N.T.

C.E. must not be used in lyddite or shellite filled shell, as dangerous chemical action may occur.

Picric powder is a mixture of 43 parts of ammonium picrate with 57 parts of saltpetre. It is used as an exploder for shellite and may also be found with certain lyddite shell.

It absorbs moisture readily and when damp is useless as an exploder.

Picric powder exploders are used in conjunction with a non-detonating fuze.

(c) *Fillings* or bursting charges consist of high explosives, viz. lyddite, shellite, T.N.T. and amatol.

Lyddite is picric acid melted and poured into the shell, where it solidifies. It is unaffected by moisture or by temperature, and consequently can be stored under almost any climatic conditions. It is a very powerful explosive and can be easily detonated. It is not used so much now as formerly, owing principally to its susceptibility to sympathetic detonation and to its tendency to form dangerous metallic picrates if brought into contact with metals, more particularly lead. Lead picrate is extremely unstable and dangerous ; for this reason all components of lyddite shell, such as fuzes, plugs, varnish, etc., must be " lead free." Special care must be taken when fuzing lyddite shell in exposed positions to keep out dust dirt, etc., which may contain lead, even in small quantities. The detonation of lyddite is indicated by dense smoke of a dark colour. Any yellow discoloration in the smoke indicates imperfect detonation.

Shellite is a modified form of lyddite, being less sensitive and not so liable to sympathetic detonation. It has a lower melting point, but is also susceptible to lead.

Trotyl or *trinitrotoluene* is known as T.N.T. It is manufactured from a derivative of coal tar and is slightly less powerful than lyddite. It is unaffected by moisture or by temperature when pure. Shell filled with T.N.T. are not so liable to accidental detonation as lyddite. Pure T.N.T. has little or no action on metals. This explosive, especially if not pure, in contact with certain metallic compounds, such as lead oxide, may form sensitive substances analogous to the picrates ; " lead free " conditions, although not so important as in the case of picric acid, should therefore be observed. It is usually melted and poured into the shell, where it solidifies. T.N.T. in crystal form is used in exploders for lyddite, T.N.T. and certain amatol-filled shell.

Amatol is formed by the mixture of T.N.T. and ammonium nitrate, in varying proportions, such as 40/60, 50/50, 80/20, the first figure in each case being the amount of ammonium nitrate used. The first two mixtures are usually poured into the shell, the T.N.T. being melted first and the nitrate stirred in. The 80/20 mixture is inserted in a dry or plastic state under pressure.

Amatol is more powerful but not so violent as lyddite, and is more difficult to detonate than either lyddite or T.N.T. It is very susceptible to moisture and should therefore be kept dry.

Ammonium nitrate is itself an explosive, and amatol should not be regarded merely as diluted T.N.T.

If crude T.N.T. is used in the making of amatol and a storage temperature of 70 degrees F. or over is experienced, some of the oils of the T.N.T. exude from the shell at the fuze-hole, etc. Where the percentage of T.N.T. is small, this exudation will not be found troublesome, particularly where exploders of C.E. are used in lieu of T.N.T. crystals. If the exudation dries in the threads of base plugs or fuze-hole bushes, there is a possibility of prematures occurring at the moment of firing. On the other hand, the most likely trouble is the liquid exudation getting at the T.N.T. in the exploder bags, which thereby may be rendered useless, causing a blind. Amatol-filled projectiles produce little smoke and, when observation is necessary, require the addition of a smoke mixture.

3. MISCELLANEOUS

Gunpowder, although no longer used as a propellant, is used in many components, such as fuzes, tubes, igniters, etc.

It is a mixture of charcoal, saltpetre and sulphur in the proportions respectively of 15, 75, 10. Many sizes of this have been manufactured, but present supply is confined to the sizes suitable for filling tubes, primers, igniters for propelling charges, magazines of non-detonating fuzes, blank charges and bursting charges of shrapnel shell and certain mortar bombs.

It is very susceptible to friction, but is unaffected by ordinary temperatures, is easily ignited and invariably explodes when ignited, giving a large flame. For this reason gunpowder is much used to ignite other explosives. As moisture or damp affects gunpowder greatly, rendering it useless, all articles containing gunpowder must be specially protected against moisture.

EFFECTS OF CLIMATE ON AMMUNITION

Continued high temperature is harmful to ammunition, which should also be shielded from the direct rays of the sun. There should be free circulation of air through a stack of ammunition.

Moisture is harmful, as it may cause steel components to rust, and because it may penetrate an absorbent component of the ammunition. Detonators are also adversely affected by damp storage.

CHAPTER I

ORDNANCE, M.L. 3-INCH MORTAR, MARK II

PARTICULARS

Material	Steel.
Weight (approximate) complete	1 qr. 14 lb.
Length without breech piece	48 inches.
Length with breech piece	51·1 inches.
Bore—	
Length from muzzle to front of striker stud	45·25 inches.
Diameter	3·209 inches.
Firing mechanism	Percussion.

CONSTRUCTION

The *mortar* (Fig. 1) consists of a barrel, breech piece, striker stud and washer.

The *Mark I barrel*, of steel, is reduced in diameter at one end and is screw-threaded on the exterior to take the breech piece with striker stud.

A hole is drilled in the outside of the barrel at 24½ inches from the muzzle to take the locating piece of the recoil spring band, and two holes at 15 inches from the muzzle to take the locating screws of the recoil stop band.

A clinometer plane and sight line are prepared on the top of the barrel towards the front end, the sight line being filled in with luminous paint.

The nature and *Mark* of the mortar, manufacturer's initials, *Mark* and year of manufacture of the barrel are stamped on the breech end.

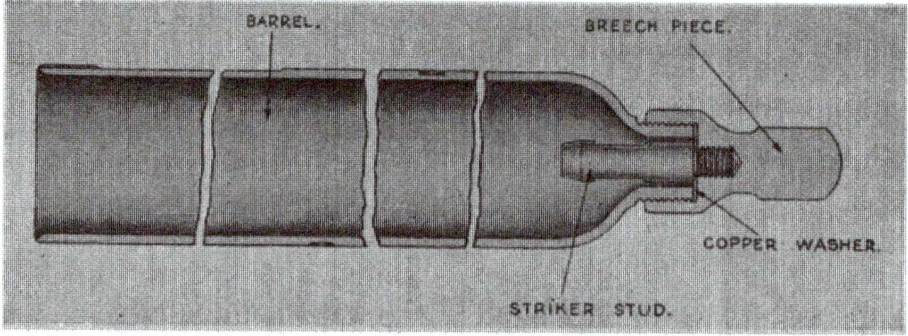

FIG. 1

The *Mark IA* barrel is a conversion of the barrel used with the *Mark I* mortar.

The *breech piece*, of steel, is bored and threaded at the front end to screw onto the threaded portion of the barrel and to receive the striker stud.   The rear end is ball-shaped to suit the socket of the base plate, flats being cut on the ball to accommodate a wrench when assembling or dismantling.

The nature and *Mark* of mortars, manufacturer's initials, *Mark* of breech piece and year of supply are stamped on the breech piece.

The *striker stud*, of steel, is cylindrical in form, the front end being coned and the rear end being screw-threaded to screw into the breech piece.   When assembled, it projects axially into the bore. Flats are formed on it to facilitate its insertion into and removal from the breech piece.

The nature and *Mark* of the mortar, manufacturer's initials, *Mark* and year of manufacture of the striker stud are stamped on the rear end.

The *washer*, of copper, is assembled between the breech piece and striker stud.   It provides a tight joint between the barrel and breech piece when the latter is in position.

CHAPTER II

MOUNTING, 3-INCH MORTAR, MARK I

(Plates 1 and 2)

GENERAL DESCRIPTION

The mounting is designed to form a firing support for the front part of the mortar barrel, the breech end of the barrel fitting into a base plate which takes the shock of firing.

An elevating gear is provided, also a traversing gear to admit of a limited amount of traverse being given without moving the legs of the mounting.

A sight is provided for laying the mortar.

The principal parts of the mounting are as follows :—

> Bipod.
> Cradle.
> Elevating gear.
> Traversing gear.
> Recoil spring.
> Recoil spring band.
> Recoil stop band.
> Buffer ring.
> Sighting gear.

BIPOD

The *bipod* (Fig. 2) consists of two tubular aluminium alloy legs, right and left respectively, each having a spiked foot and stay bracket riveted to it. The legs are hinged at the upper end to a crosshead and, towards the bottom, are provided with stays, the inner ends of which are connected to the elevating screw tube when the mounting is assembled.

Each stay is in two parts which are held rigid by a locking pin with nut, spring washer and keep pin when the mounting is in action.

The *crosshead* consists of two steel links, front and rear, connected together by bolts with distance tubes. The front link has a toothed arc on its front face to engage the teeth on a clamping plate actuated by a clamping handle. This arrangement permits of the mounting being levelled to counteract any difference due to the legs of the bipod not being on the same level.

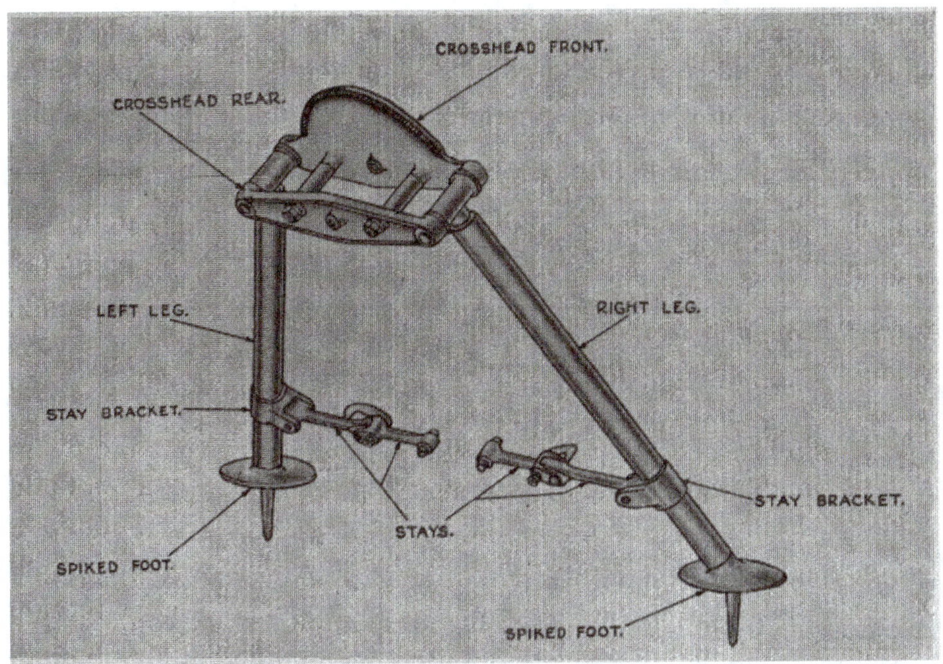

FIG. 2

CRADLE

The *Mark II cradle* consists of a body and sight supporting bracket.

The *body* (Fig. 3), of manganese bronze, is circular in shape to

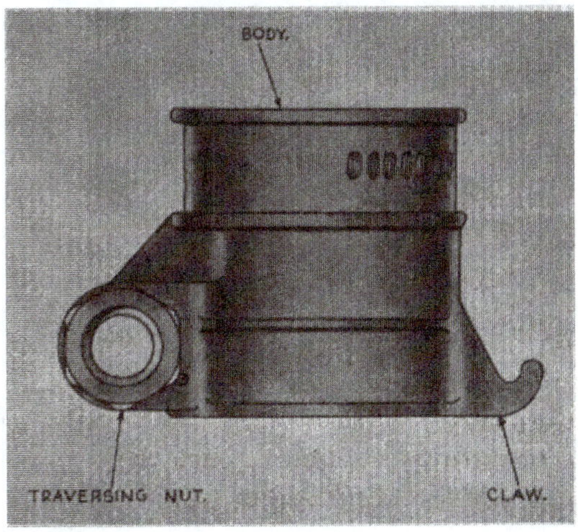

FIG. 3

receive the barrel of the mortar. A traversing nut is formed on the underside and is fitted with a screwed bush to receive the traversing screw. A claw formed on the upperside, towards the rear, takes the toggle catch of the recoil spring.

The body is prepared with teeth around a portion of its periphery at the front to engage the teeth of a cross-levelling worm spindle.

The *sight supporting bracket* (Fig. 4), of steel, consists of a band in two parts, hinged together, clamped to the front of the cradle

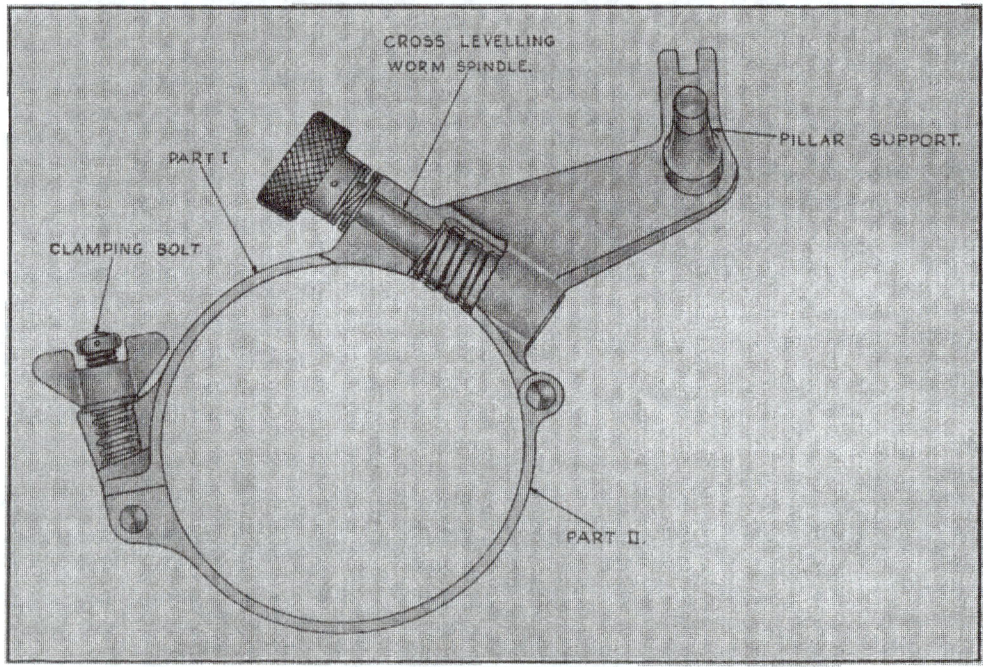

Fig. 4

body by means of a clamping bolt with wing nut, spiral spring and washer.

Part I of the bracket is formed with a pillar support for the sight and a bearing for the cross-levelling worm spindle. The latter gears with the teeth on the cradle body and is employed in conjunction with the cross-level bubble of the sight to enable the sight to be aligned in the vertical plane of the barrel. When the bubble is brought into the centre of its run, through the cross-levelling worm spindle, the sight should be vertical.

The *Mark I cradle* differs from the *Mark II* in not being fitted with a screwed bush in the traversing nut.

Elevating Gear

The *elevating gear* (Fig. 5) consists of an elevating bracket, elevating screw, bevel pinion with operating handle, clamping arrangement and an elevating screw tube.

The *elevating bracket* consists of a tubular steel body enlarged at

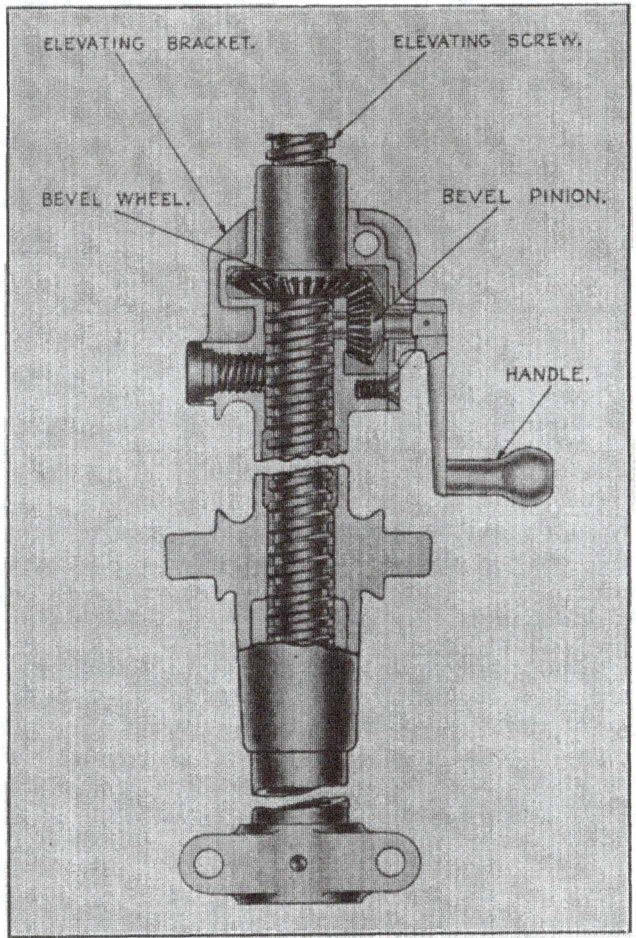

Fig. 5

its lower end for the reception of the elevating screw tube and is provided with trunnions near the centre for attachment to the front and rear links of the crosshead. The upper end is enlarged and recessed internally to receive a bevel wheel, a boss on the latter being screw-threaded internally to receive the elevating screw. Externally, the upper end of the bracket is cut away and recessed to receive a bevel pinion, the recess being provided with an aluminium alloy cover attached to lugs formed on the bracket. The cover is bored centrally and bushed to receive the spindle of the bevel pinion and is retained in the closed position by a securing screw.

A boss formed on the bracket, opposite the cut-away portion, is bored and screw-threaded to accommodate a clamping screw and is recessed for a clamping plate and spring.

The *elevating screw* consists of a steel rod screw-threaded for the greater portion of its length to engage the boss on the bevel wheel. The upper end is attached to the yoke of the traversing gear and a stop is screwed to the lower end to prevent the screw over-running the boss of the bevel wheel.

The *bevel pinion* engages the bevel wheel on the elevating screw and is operated by an operating handle secured by a keep pin to the outer end of the bevel pinion spindle.

When the handle is turned, the bevel pinion, gearing with the bevel wheel, causes the latter to revolve. The boss of the bevel wheel engaging the threads of the elevating screw causes the screw to move through the boss, thus bringing the mortar to the required elevation.

The *clamping arrangement* consists of a steel clamping screw threaded at both ends and left plain about the centre. The inner end of the clamping screw is screwed into the boss of the elevating bracket and is riveted in position. The plain portion receives a steel spiral spring, clamping plate and guard plate. The outer end of the screw has a cranked clamping handle screw-threaded to suit the clamping screw and retained by a screwed collar and keep pin. The lower surface of the clamping plate is formed with teeth to engage the arc on the cross-head of the bipod.

When the clamping handle is screwed inwards, the guard plate pressing against the clamping plate causes the teeth on the latter to engage with those on the arc, thus clamping the mounting at the required level. When the handle is released the spring acting between the inner end of its recess and the clamping plate causes the latter to become disengaged from the arc.

The *elevating screw tube*, of aluminium alloy, is riveted to the lower end of the elevating bracket. Brackets are secured to the lower end of the tube for attachment to the stays of the bipod legs.

Traversing Gear

The *Mark II traversing gear* (Fig. 6) consists of a yoke, traversing screw and handle.

The *yoke*, of steel, is bow-shaped and is formed with bearings at its extremities to receive the traversing screw. It is bored through the centre for attachment to the upper end of the elevating screw.

The *traversing screw* consists of a steel rod screw-threaded for nearly the whole of its length, being left plain at the ends to enter the bearings in the yoke. The left end projects through its bearing and is prepared to receive a cranked handle retained in position by a castle nut and keep pin.

When the handle is operated, the screw is revolved through a traversing nut on the underside of the cradle, causing the latter to move along the traversing screw.

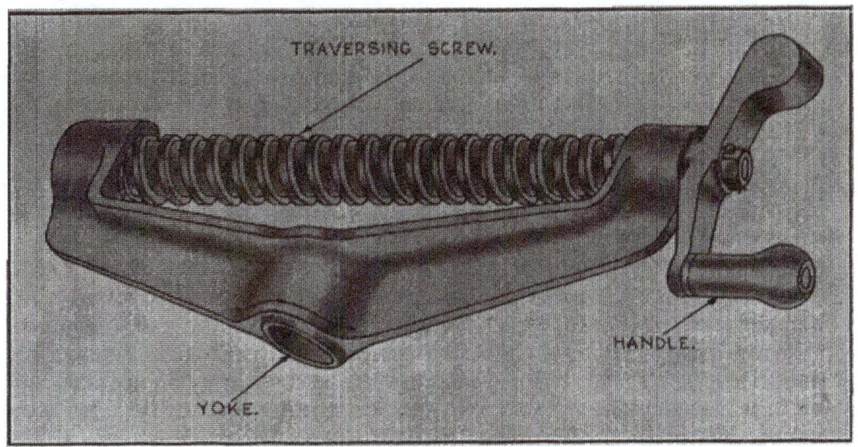

Fig. 6

The *Mark I traversing gear* differs from the *Mark II* in the traversing screw being in two pieces, the screw being keyed to a shaft. The left end of the shaft is formed with a square projection to receive the handle, the latter being retained on the shaft by a taper pin and a keep pin chained to the yoke.

Recoil Spring

The *recoil spring* (Fig. 7), of steel, is spiral in form. It is furnished, at its lower end, with an eye to engage a hook on the recoil spring band and, at its upper end, with a hook and toggle catch formed integral. The toggle catch engages the claw formed

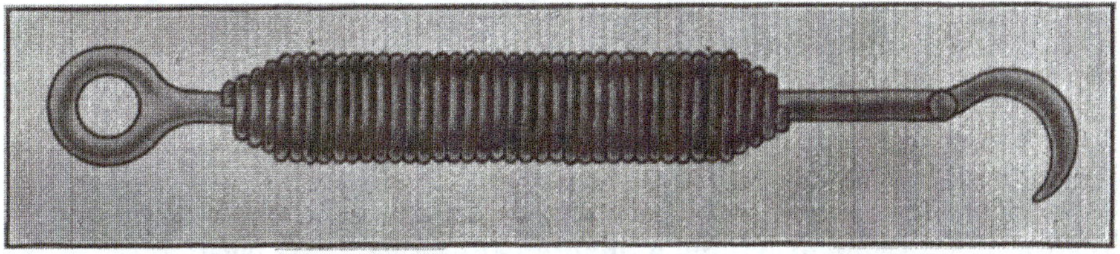

Fig. 7

on the cradle body and secures the cradle to the mortar when assembled for action. When the mortar is fired, the cradle has a tendency to jump forward on the barrel. The function of the spring is to return the cradle to its normal position on the barrel.

When the mortar and mounting are dismantled for transporting purposes, the toggle catch is disengaged from the cradle body and attached to a claw formed on the recoil stop band.

RECOIL SPRING BAND

The *recoil spring band*, of steel, is clamped around the mortar approximately about half-way along the barrel. A hook is formed on the band to engage the eye on the lower end of the recoil spring.

The band is positioned on the barrel by means of a locating piece engaging a hole drilled in the mortar barrel.

RECOIL STOP BAND

The *recoil stop band*, of manganese bronze, is clamped around the barrel below the cradle body. A claw is formed on the band to take the toggle catch of the recoil spring when the mortar is not in position and two bosses on the band are screw-threaded to accommodate locating screws, the latter engaging holes drilled in the mortar barrel.

BUFFER RING

The *buffer ring*, of hemp rope, is placed around the barrel between the lower edge of the cradle body and the recoil stop band to prevent metal to metal contact, and to receive the blow of the cradle when returned to its normal position on the barrel after firing.

SIGHTING GEAR

The *sighting gear* is attached to the mounting by means of the sight supporting bracket on the cradle. It consists of a sight of the reciprocating type with a clinometer incorporated, the former providing an all-round field of view for indirect laying for line and the means of cross-levelling the sight, and the latter enabling the mortar to be layed at the required quadrant elevation.

SIGHT, 3-INCH MORTAR, MARK I

(Plate 3)

The *sight* consists of a range scale bracket, worm wheel bracket, collimator bracket and range scale slider.

The *range scale bracket*, of manganese bronze, consists of a body in the form of a segment having a dovetailed groove on its periphery for the reception of a range scale strip. A semi-circular projection on the lower side of the body is recessed to fit over the supporting pillar on the sight supporting bracket and is bored transversely for the reception of a locking pin. The projection is engraved with a direction arrow and the words TO LOCK, the engraving being filled in with black wax.

The *locking pin* consists of a spindle with latch, the spindle being cut away to engage the supporting pillar when the sight is in position on the mounting. The latch is provided with a spiral spring and hardened steel ball, the latter engaging in one of two holes on the outer face of the semi-circular projection according to whether the spindle is in the locked or unlocked position.

The *range scale strip* is engraved with two yard scales, one for full charge and the other for reduced charge. The full charge scale is graduated in yards from 525 to 1,600 in multiples of 25 yards, and the reduced charge scale from 275 to 850, also in multiples of 25 yards.

The *worm wheel bracket* consists of a manganese bronze body with a worm wheel and an aluminium alloy deflection dial.

The *body* is circular in shape with two lugs formed on its upper surface for the attachment of a collimator bracket. It is bored vertically through the centre for the worm wheel pivot and has three slotted holes for the attachment of the deflection dial. Two slotted holes are also provided for the attachment of the worm wheel.

The *worm wheel* is secured to the underside of the body and is pivoted to the range scale slider to permit of all-round movement of the collimator bracket for the purpose of " laying back " if necessary.

The *deflection dial* is secured to the body by three screws working in the slotted holes which permit of adjustment of the dial. It is graduated from 0 to 180 degrees Right and Left in multiples of ten degrees, the right graduations being marked with the letter R and the left with the letter L to signify right and left deflection respectively. The engraving is filled in with black wax.

The *collimator bracket* consists of a manganese bronze body, formed with fore and hind sights. It is hinged at the front to the lugs on the worm wheel bracket, the hinge pin being fitted with a spiral spring to take up play. The rear end of the bracket is provided with an adjusting screw having a knurled head. The lower end of the adjusting screw bears on the top surface of the worm wheel bracket and provides the means for adjusting the line of sight in the vertical plane. Two screws are provided for the attachment of the No. 1 collimator.

The *range scale slider* consists of a manganese bronze or aluminium alloy body, deflection worm spindle bearing and spring plunger, deflection worm spindle, deflection clicker and steel clamping bolt.

The *body* is in the form of a segment. It is formed with a circular recess at the front end to receive the worm wheel and bracket, a boss formed in the centre of the recess being bored and threaded to receive a steel pivot for the worm wheel. A projection on the left side of the recess has a socket for the reception of a longitudinal level employed in conjunction with the elevating gear of the mounting. When the bubble is brought into the centre of its run by operating the elevating gear the mortar is layed at the quadrant elevation applied to the sight.

The underside of the body is dovetailed to form a slide and fits over the range scale bracket, being clamped to the latter by the clamping bolt. A transverse projection about midway on top of the body has a socket for a cross-level bubble employed with the worm spindle on the sight supporting bracket of the cradle to compensate for difference in the level of the legs. A lug is formed on the body, above the transverse projection, for the attachment of the deflection worm spindle bearing.

An adjustable reader is secured to the lower end of the body by screws. Two arrow heads on the reader provide the means of reading the scales. The reader is dovetailed on the underside to slide along a transverse bar marked FULL and RED, a clamping screw with a knurled head clamping the reader in the required position.

The *deflection worm spindle bearing*, of aluminium alloy, is bored transversely to receive the worm spindle and is formed with two lugs on the underside for attachment to the range scale slider body by means of a hinge pin. A recess in the underside forms a seating for a plunger and spiral spring to keep the worm spindle engaged with the worm wheel. The spring plunger also provides a means for throwing the spindle out of gear with the worm wheel for quick and approximate settings of large deflection angles by hand, the worm spindle being employed for deliberate and final setting. A thumb piece formed on the rear of the spindle bearing enables the bearing to be depressed against the spring plunger when disengaging the worm spindle from the worm. A slot in the thumb piece accommodates the upper end of a spring catch screwed to the range scale slider body. A stop piece, riveted to the rear face of the spring catch, engages the underside of the thumb piece and prevents the worm spindle from becoming disengaged from the worm wheel on shock of discharge.

An arrow is engraved in the centre of the bearing, towards the front, for reading the scale on the deflection dial and an arrow on each side, marked R and L respectively, act as readers for the deflection drums of the deflection worm spindle. The arrows and letters are filled in with black wax.

The *deflection worm spindle*, of rolled bronze, is formed with a thread to engage the worm wheel of the range scale slider body. The right end of the spindle is enlarged to form the right deflection drum and the left end is screw-threaded to receive the left deflection drum after the spindle has been assembled in the deflection worm spindle bearing. The drums are graduated from 0 to 10 degrees in multiples of a degree, subdivided in multiples of 30 minutes. They are engraved with the letters L and R, to signify left and right deflection respectively, the graduations and letters being filled in with black wax. The drums are knurled on their outer edges.

The inner surface of the right deflection drum is prepared with

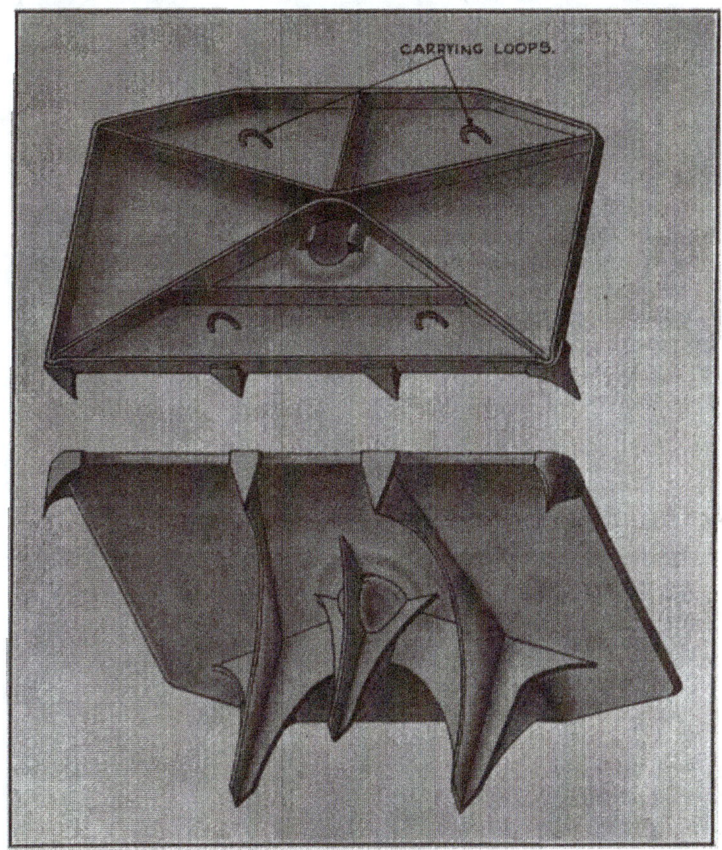

FIG. 8

a circle of twenty equally spaced holes to engage with the ball of the deflection clicker. Provision is also made on the spindle for the use of a spring and plain washer to take up backlash.

The spindle is retained in its bearing by means of two nuts which also secure the left deflection drum to the spindle. A plain spring washer on the worm spindle, interposed between the left deflection drum and the worm spindle bearing, takes up any backlash.

The *deflection clicker* consists of a steel ball and spiral spring housed in a recess in the right side of the deflection worm spindle bearing in a position to enable the ball to engage the holes in the right deflection drum. When the drum is revolved to apply deflection to the sight, each successive engagement of the ball with a hole in the drum produces an audible click which represents an angular movement of 30 minutes on the deflection drums.

The *clamping bolt*, of steel, is provided with a wing nut to clamp the range scale slider in any required position on the range scale bracket.

PLATE, BASE, 3-INCH MORTAR

The *base plate* (Fig. 8) is of steel, with stiffening plates welded to its upper surface for strengthening purposes. The top of the plate is provided with four carrying loops and is prepared with a recess to take the ball-shaped portion of the breech piece. The recess is formed with two lugs to act as a locking device, a quarter of a circle turning movement of the mortar being necessary to engage or disengage the breech piece.

Four spiked plates welded to the front edge of the base plate and stiffening plates welded to the under surface and forming spikes serve to ensure a firm anchorage.

TABLE OF WEIGHTS AND DIMENSIONS

DIMENSIONS

	Inches
Mortar—	
Length, overall	51·1
Diameter (external)	3·7
Mounting—	
Length, with sight removed and cradle swung over	36
Width, legs spread	28
Width, legs folded	10·5
Width, over traversing gear	14·25
Depth, over traversing gear	8·125
Base plate—	
Length	22·5
Width	14·25
Depth (over spikes)	9

WEIGHTS

	lb.	oz.
Mortar (including spring and two bands attached)	44	0
Mounting (including cradle)	44	9
Base plate	37	0

CHAPTER III

MISCELLANEOUS STORES

Apparatus, Practising Loading, M.L. 3-inch Mark II Mortar

The *apparatus* consists of a *Mark I* barrel converted for use in conjunction with the 3-inch mortar mounting and base plate for the purpose of practising loading.

A portion of the upper side of the barrel near the breech end is cut away to provide an exit for the drill bomb. A hard wood block, faced with a steel plate, is fitted to the barrel adjacent the opening to divert the bomb from the vicinity of the base plate.

A breech piece and copper washer are fitted to the rear end of the barrel and the latter is arranged for the attachment of the recoil spring band, recoil stop band and cradle of the mounting.

Preparation for use

To prepare the M.L. 3-inch mortar equipment for use as an apparatus, practising loading, remove the sight and barrel.

Remove the recoil stop band and recoil spring band from the service barrel and assemble these, with the buffer ring, on the apparatus, practising loading barrel. Insert this barrel in the cradle and base plate.

A mat should be used to avoid damage to the base plate and the drill bombs.

Bag, Tool, 3-inch Mortar

The *tool bag*, in the form of a satchel, is made of canvas and is provided with a webbing shoulder sling. The interior of the bag is furnished with four pockets and it is closed by a flap secured by a strap and buckle. The sling is in two parts, one part having a brass buckle for adjustment purposes and the other having a brass tag.

Carrier, Ammunition, M.L. 3-inch Mortar, No. 1

The *ammunition carrier* consists of three brown rolled paper tubes secured together by steel bands tightened by iron wire passed between the tubes and around the bands, hardwood spacers, around which the iron wire passes, being placed between the tubes.

The tubes are sized and shellac varnished inside and out and are closed at both ends by tinned-plate caps, the nose end caps having a rolled paper cylinder glued to the inside to form a cavity for the fuze, the caps being secured to the tubes with shellac. The base end caps form the lids and are strengthened by a millboard disc glued to the inside.

A cotton webbing securing band passes around the tubes, through guides soldered to the end caps, and is secured on top by a buckle. A cotton webbing handle is provided for carrying purposes.

CARRIER, AMMUNITION, M.L. 3-INCH MORTAR, DUMMY

The *dummy ammunition carrier* consists of a wood block, approximately 18 in. × 11½ in. × 4 in., having a lead or cast iron block let in to give a total weight of 33 lb.

A tarred hemp rope handle is provided, the ends being passed through holes in the block and knotted to secure the handle to the block.

CASE, MORTAR CLINOMETER

The *case* is of leather and is shaped to take the clinometer. It is fitted with loops to pass over the waist belt.

CASE, 3-INCH MORTAR SIGHT

The *case* is of leather and is provided with a shoulder strap for carrying purposes. It is designed to house the sight for transport or storage.

CLINOMETER, FIELD

The *field clinometer* is designed to permit of its being quickly set to any required graduation.

The *Mark VI clinometer* (Fig. 9) consists of the following parts :

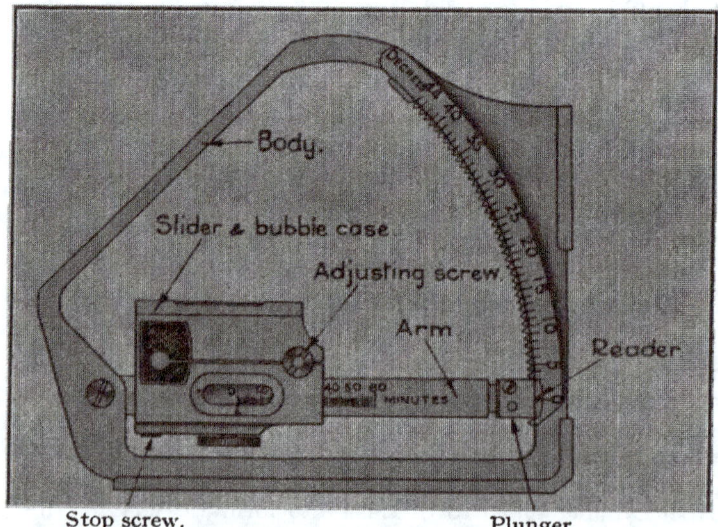

Fig. 9

body, arm, plunger, slider and bubble case—all being made of gun-metal.

The *body* has two plane surfaces at right angles to each other. A transverse hole is bored for the pivot of the arm, opposite which is a toothed arc, each tooth representing one degree. The sides of the teeth are graduated in degrees on one side from 0 to 44 degrees, on the other from 45 to 89 degrees, an arrow on the side of the plunger acting as an index. The surface forming the base of the clinometer is marked with an arrow to indicate the direction of the target, and with the words USE THIS BASE FOR 0° TO 45°. The second surface has an arrow indicating direction of target, together with the instruction 45 DEGREES TO 90 DEGREES.

The upper portion of the body is slotted longitudinally to permit of observation of the bubble when the arm is set at angles between 35 and 55 degrees.

The *arm*, along which moves the slider with bubble, is pivoted to the body. It is slightly curved so that, as the slider moves away from the pivot, the axis of the bubble is placed at an increasing angle to the base of the clinometer, the total movement being equivalent to one degree. The arm is graduated on each side from 0 to 60 minutes, reading in opposite directions, and is bored from the outer end to take a plunger and spring. A small slot on the under surface, in conjunction with the stop screw of the slider, limits the movement of the latter.

The *plunger* has a spindle and head in one piece. The spindle portion fits in the arm against a spiral spring. The head is rectangular, with teeth to engage the arc on the body. Side plates are fixed to the head, on each of which an index is engraved for the degree scale. The plates are roughened and project on each side of the arc.

The *slider and bubble case* is rectangular in shape and divided into upper and lower portions by a saw-cut. The solid end is towards the pivot of the arm and the open ends are connected by adjusting and clamping screws. The upper portion is bored longitudinally for the " bubble, spirit, glass, C," the upper surface being partly cut away to expose the lines on the bubble. The ends are closed by caps. The lower portion is prepared to slide along the arm and has slots on each side to expose the minute graduations, which are read by an arrow. A clamping screw, with milled head, and a stop screw are fitted to the underside, whilst a flat nickel silver spring lies between the slider and arm. The slider can be clamped to any position on the arm by the milled headed clamping screw.

The graduations on the clinometer are filled in with black wax, the body is nickel plated in parts and enamelled black elsewhere. The instrument is carried in a wooden box.

The *Mark V* differs from the *Mark VI* in the form of slider and bubble case. The case is a separate fitting and is secured in the slider by end caps. Three adjusting screws support and secure the bubble case at its inner end.

The *Mark IV* differs from the *Mark V* in the graduation of the arc and slider, these being identical on both sides, i.e. 0 to 44 degrees on the arc and 0 to 60 minutes on the arm. No instructions are engraved as to how to place the clinometer on the plane.

The *Mark III* differs from the *Mark IV* in the method of attaching the bubble casing to the slider. The case is pivoted on trunnions at one end and is supported on a capstan-headed screw at the other, the latter providing the means of adjustment. The upper portion of the body is not slotted.

In all *Marks* suitable provision is made for effecting the necessary adjustments for index error.

To set the clinometer

Press in the plunger and move the arm until the reader on the plunger is opposite the required graduation on the degree scale. Release the plunger, unclamp the clamping screw and move the slider until the arrow on it is opposite the required graduation on the minute scale. Clamp the slider. The degree and minute scales on the same side of the clinometer must always be used in conjunction.

To use the clinometer

Great care should be taken to see that the clinometer plane on the mortar is clean and free from grit, paint, etc., and that the field clinometer is in correct adjustment for index error, also that the bubble axis remains parallel to the longer axis. The instrument must be placed in the same position each time and must be correctly set, the elevation being adjusted so that the bubble is central each time.

CLINOMETER, MORTAR

The *Mark III clinometer* (Fig. 10) consists principally of an arc, bubble slider and spring, bubble case and bubble.

The *arc*, of duralumin, is prepared with a groove to take a slider with spring, a stop screw at each end of the groove preventing the slider from over-running the arc. A scale, graduated from 0 to 90 degrees, is engraved on one side of the arc, the graduations being read by means of an arrow engraved on the slider. An arrow, indicating the direction of the target, and the word TARGET is engraved on both sides of the arc. All markings are filled in with black wax.

The underside of the arc has a dovetailed groove to receive a steel base secured in position by screws.

The *slider*, of gunmetal, is prepared on its upper surface to receive a brass bubble case with a No. 1 glass spirit bubble, the case being secured to the slider by four screws. The slider is kept up to its work in the groove in the arc by means of a steel spring.

When not in use the clinometer must be kept in its case.

The *Mark II clinometer* differs from the *Mark III* in not having a removable steel base.

COLLIMATOR, No. 1

The *No. 1 collimator* consists of a metal body in which are mounted a lens and wedge-shaped window. At one end of the lens an arrow point is marked which, being in the focal plane of the lens, will, when seen through the lens, appear to be at an infinite distance.

The wedge-shaped window transmits light into the lens, thus illuminating the arrow point.

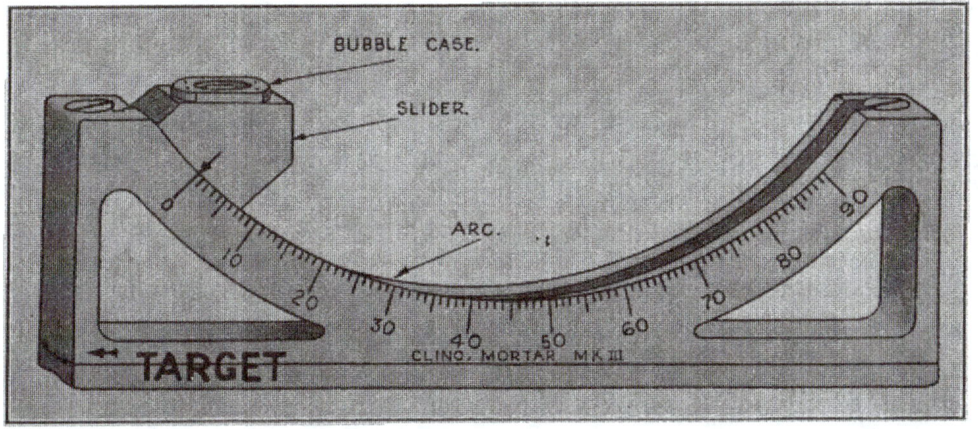

Fig. 10

COVER, MUZZLE, 3-INCH MORTAR

The *Mark III muzzle cover* is made of leather and is suitably shaped to fit the muzzle of the mortar. It is provided with an attachment strap with a buckle and a running loop for adjustment purposes. The loop has a quick-release link by means of which the strap is connected to the No. 1 harness when in man transport, or to the recoil spring catch when the harness is not in use. A leather handle is stitched to the front of the cover to enable it to be withdrawn from off the muzzle of the mortar.

The *Mark II cover* is made of canvas with a wire ring and has a securing cord.

The *Mark II* cover* is the *Mark II* modified to approximate to the *Mark III* by the provision of an attachment strap for use with the No. 1 harness.

Harness, Mortar

The *harness* is made of webbing with leather strengthening pieces and straps. The shoulder pads are made of sponge rubber covered with dowlas.

The *No. 1 harness* is designed to carry the barrel and consists of a shoulder pad and sling in two parts. Part I comprises a strap with buckle for attachment to the barrel, an attachment strap and a muzzle cover strap with quick-release link. Part II comprises a strap with breech piece ring and buckle.

The *No. 2 harness* is for carrying the base plate and comprises a shoulder pad, base-plate pad and sling in two parts. Part I is provided with a spring swivel hook tab. Part II has a loop, spring swivel hook tab and buckle.

The *No. 3 harness* is for carrying the mounting and comprises a shoulder pad and sling in two parts. Part I consists of a sling strap with 2-inch ring and attachment strap with spring swivel hook. Part II consists of a buckling piece with buckle, a $2\frac{1}{2}$-inch ring and two leg spike straps each with a 2-inch ring.

Implements, Ammunition

Extractor, cartridge, M.L. 3-inch mortar.—The extractor, of steel, is made in the form of pincers. The jaws are shaped to suit the primary cartridge, the closing movement being limited by means of a bar secured to the inside of one of the handles.

Key, No. 119.—The key consists of a length of steel rod with a projection formed at one end, the other end being flattened and having a hole drilled through it. It is for use in removing or inserting No. 138 fuzes.

Posts, Aiming, B.C. and 3-inch Mortar, Mark I

The *aiming post* consists of a steel tube jointed about its centre through a link and sleeve joint, and is fitted with a diamond-shaped head. A steel spike is fitted to the lower end of the tube and a hinged step is provided for the foot of the person planting the post.

The link and sleeve joint enables the aiming post to be folded for convenient carriage.

Two aiming posts are required for use with each mortar. When used at night, an adjustable lamp bracket to carry an electric lamp is attached to one of the aiming posts.

Rod, Cleaning

The *cleaning rod* consists of a tubular steel rod with a cross-handle and steel wad hook. The wad hook screws into a coupling socket at the end of the steel rod and is secured by rivets.

Scrubber, Brass

The *scrubber* consists of 6 oz. of brass turnings and is used, in conjunction with the cleaning rod, for cleaning the bore of the mortar barrel.

Wrench, Breech Piece, M.L. 3-inch Mark II Mortar

The *wrench*, of steel, consists of a flat bar having a circular projection at one end shaped to fit over the breech piece of the mortar.

CHAPTER IV

Statement of Ammunition

Nature of bomb	Mk.	Nature of filling or charging	Nature and weight of exploder	Fuze	Cartridges		Weight								
					Primary	Secondary	Filling			Fuze			Filled bomb		
							lb.	oz.	dr.	lb.	oz.	dr.	lb.	oz.	dr.
High ex-plosive.	II	Amatol 80/20	6¼-dr. C.E. pellet.	No. 138	95-gr. bal-listite.	100-gr. or 93-gr.* N.C.(Y) (6 to set)	1	2	4	1	4	8	10	0	10
			1-oz. 4-dr. C.E. Pellet.	No. 150						0	10	4	9	7	2
Smoke.	I	Phosphorus.	6¼-dr. C.E. pellet.	No. 138	,,	,,	1	7	5	1	4	8	10	5	11
			1-oz. 4-dr. C.E. pellet.	No. 150						0	10	4	9	12	3
Practice.	I	Ammonium alum and buckshot. Bursting charge of 7 oz. G.12 powder.	7-dr. prim-ers.	No. 150P	,,	,,	1	8	2	0	10	4	10	0	0

* The 93-gr. N.C.(Y) augmenting cartridge is obsolescent.

AMMUNITION

GENERAL REMARKS

Ammunition for the M.L. 3-inch mortar is normally issued in complete rounds, except for overseas, and consists of a cartridge with striker, six augmenting cartridges, bomb and fuze.

It is packed three rounds, complete with cartridges and fuzes in position, in a No. 1 carrier, two carriers to a box, the tail unit being fitted with a tinned-plate protecting cup and a waterproof all-wool cloth cover.

For overseas, the bombs and fuzes are packed separately to the cartridges when fuzes without shutters are used.

Smoke and practice bombs for use with unshuttered fuzes are issued plugged, six in a box, the cartridges and fuzes being packed separately.

Details of certain identification stampings and stencillings on the ammunition and packages will be found in Chapter V. The table on page 34 shows the ammunition approved for use with this equipment.

CARTRIDGES

CARTRIDGE, M.L. MORTAR, 95-GRAINS BALLISTITE, WITH STRIKER CLIP, MARK VI

The *cartridge* (Fig. 11) consists of a paper case having a brass

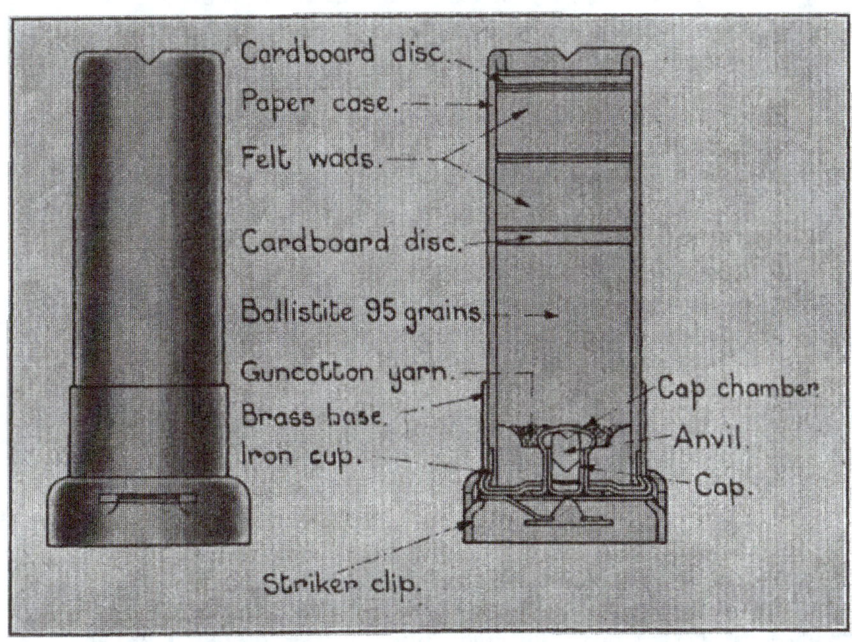

FIG. 11

base stiffened internally with an iron cup and compressed paper wad. A cap chamber is formed in the base and fitted with a brass anvil and copper cap, the latter containing about 0·5 grains of detonating composition covered with a tinfoil disc. A priming of about five grains of guncotton yarn is placed around the cap chamber.

The charge consists of 95 grains of loose ballistite. Above the charge are placed, in the order named, a cardboard disc, two felt or feltine wads and a closing disc of cardboard covered with discs of blue paper, the whole being secured in position by the top edge of the case being turned over.

The case is coated externally with shellac varnish.

The *striker clip* fits over the base of the cartridge and consists of a steel or tinned-plate guard, brass striker disc and a steel striker. The guard is cup-shaped and is secured around the edge of the base by four lips on the guard being pressed onto it after assembly, the lips also serving to secure the striker disc. A portion of the disc is cut away to form a tongue to which the striker is secured.

The *Mark V cartridge* differs from the *Mark VI* in being fitted with a different pattern striker clip, the disc being made of steel instead of brass.

The earlier *Marks* of cartridge are obsolete.

Cartridge, M.L. 3-inch Mortar, Augmenting, 100-Grains N.C.(Y), Mark II

The *cartridge* consists of 100 grains of N.C.(Y) powder contained in a cambric bag choked at the mouth with sewing silk.

The *Mark I cartridge* differs from the *Mark II* in the powder being contained in a cylindrical celluloid container, the rear end being perforated and the front end fitted with a hardwood plug.

Cartridge, M.L. 3-inch Mortar, Augmenting, 93-Grains N.C.(Y), Mark I

The cartridge consists of 93 grains of N.C.(Y) powder contained in a cambric bag choked at the mouth with sewing silk.

PROJECTILES

Bomb, M.L., H.E., 3-inch Mortar, 10-lb., Mark II

The *bomb* (Fig. 12) consists of a steel body, streamlined in shape.

When filled and fuzed with No. 138 fuze it weighs 10 lb. 0 oz. 10 dr. With a No. 150 fuze it weighs 9 lb. 7 oz. 2 dr.

The cavity tapers towards the head and base, the head being threaded to receive a nose container and the base prepared for an adapter.

Externally, the body is formed with two guide bands.

The *nose container*, of steel, is formed with a flange at the mouth to suit the contour of the bomb and is screw-threaded below the flange for a portion of its length to screw into the head of the bomb. It is screw-threaded internally at the mouth to receive the fuze or plug.

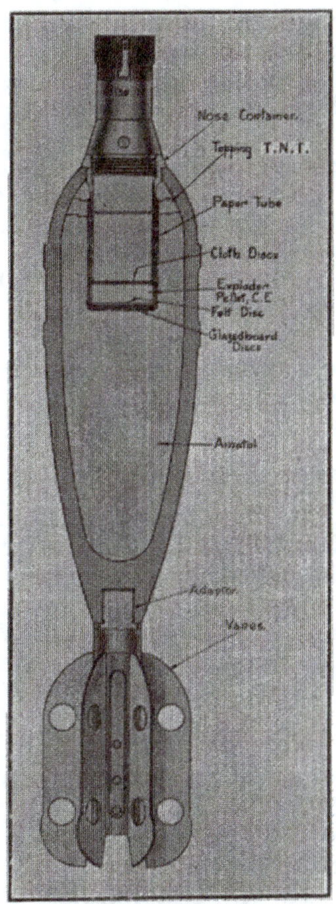

FIG. 12

The *adapter*, of steel, is screw-threaded to two diameters, the smaller diameter to screw into the base of the bomb and the larger diameter to receive a container. It is provided with a screwdriver slot to permit of a screwdriver being employed to screw it into position.

The *container*, of steel, is tubular in form and is perforated with eighteen holes to distribute the gas pressure equally around the bomb.

It is threaded internally at the front to screw onto the adapter and is secured in position by stabbing in three places. The space in the container immediately below the head of the adapter is filled with a hardwood plug secured in position by a brad. A steel spring secured in a groove in the side of the plug, is provided for retaining the 95-grain cartridge in position in the rear end of the container.

The *vanes*, six in number, are made from three pieces of steel plate suitably shaped and secured to the exterior of the container by brazing or welding. Alternatively, the vanes may be made from six pieces of steel plate secured to the container by spot welding.

The bursting charge consists of 1 lb. 2 oz. 4 dr. of 80/20 amatol with a topping of T.N.T. to assist detonation. A cavity is formed in the bursting charge to accommodate the nose container and is lined with a waxed paper tube. A 6¼-dr. C.E. exploder pellet is fitted in the nose container when a No. 138 fuze is used, and a 1-oz. 4-dr. C.E. exploder pellet when a No. 150 fuze is used.

The tail unit is provided with a waterproof all-wool cloth cover for transport and storage purposes.

BOMB, M.L., SMOKE, 3-INCH MORTAR, 10-LB., MARK I

The *smoke bomb* (Fig. 13) differs principally from the H.E. bomb in being charged with 1 lb. 7 oz. 5 dr. white phosphorus. The phosphorus is inserted through a charging hole in the side of the bomb, the hole being closed by a charging hole plug.

When a No. 138 fuze is employed, a 6¼-dr. C.E. exploder is used. When a No. 150 fuze is employed, a 1-oz. 4-dr. C.E. exploder is used.

The fuze-hole plugs for use with the smoke bomb are :—

> When prepared for a No. 138 fuze, a No. 4 plug, fuze-hole, 1·375-inch with leather washer.
> When prepared for a No. 150 fuze, a No. 5 plug, fuze-hole, 1·375-inch with leather washer.
> When filled and fuzed with No. 138 fuze, the bomb weighs 10 lb. 5 oz. 11 dr. With the No. 150 fuze it weighs 9 lb. 12 oz. 3 dr.

BOMB, M.L., PRACTICE, 3-INCH MORTAR, 10-LB., MARK I

The *practice bomb* (Fig. 14) is similar in construction to the H.E. bomb but is fitted with a nose adapter instead of a nose container. The filling consists of a quantity of ammonium alum in the bottom of the bomb with a mixture of ammonium alum and buckshot on top, the whole being retained in position by a felt disc.

The bursting charge consists of 7 oz. of G.12 gunpowder contained in a burster bag, the mouth of the bag being choked with sewing silk. Two 7-dram primers are placed on top of the burster bag as a medium for conveying the flash from the fuze to the bursting charge.

When filled and fuzed with No. 150P fuze the bomb weighs 10 lb.

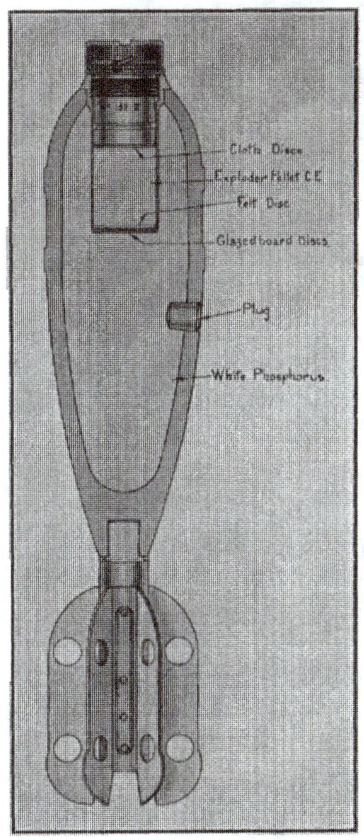

Fig. 13

The plug for use with the practice bomb, when prepared for use with the No. 150P fuze, is the No. 5 plug, fuze-hole, 1·375-inch is fitted.

Spring, Retaining, Cartridges, Augmenting, Mortar Bomb

The *spring* is employed to retain the augmenting cartridges between the vanes of the bomb. It is made of steel wire, spirally wound, with a hook formed at each end, and is passed through the holes in the vanes and around the cartridges.

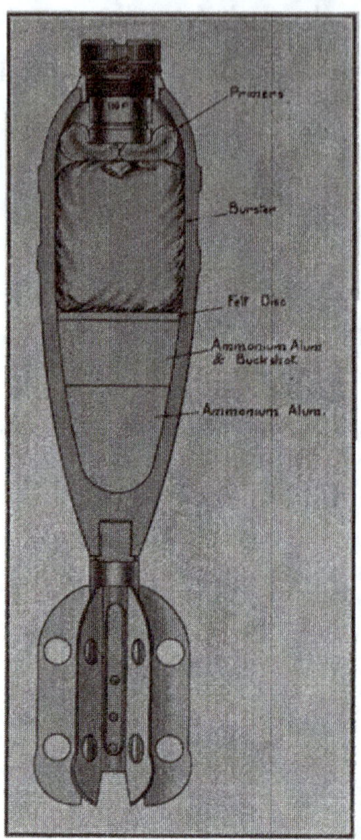

Fig. 14

FUZES

Fuze, Percussion, D.A., No. 138, Mark I

The *fuze* (Fig. 15) consists of the following principal parts :—

Body, magazine with plug, shutter with spring and detonator, guide bush, striker with needle and flanged head, striker spring, arming sleeve with arming spring, six balls, retaining sleeve, safety cap with spring and washer.

The various parts are metal except where otherwise stated.

The *body* is formed with a conical flange below which it is threaded for a portion of its length to the 1·375-inch fuze-hole gauge. It is milled near the top to engage the safety cap spring and screw-threaded externally at the bottom to receive the magazine. A keyhole is drilled in the flange for fixing purposes. The body is bored internally to accommodate the percussion and arming arrangements, and screw-threaded at the top to receive the guide bush.

The *magazine* is recessed at the top and screw-threaded internally to screw on to the body, being retained in position by a set screw.

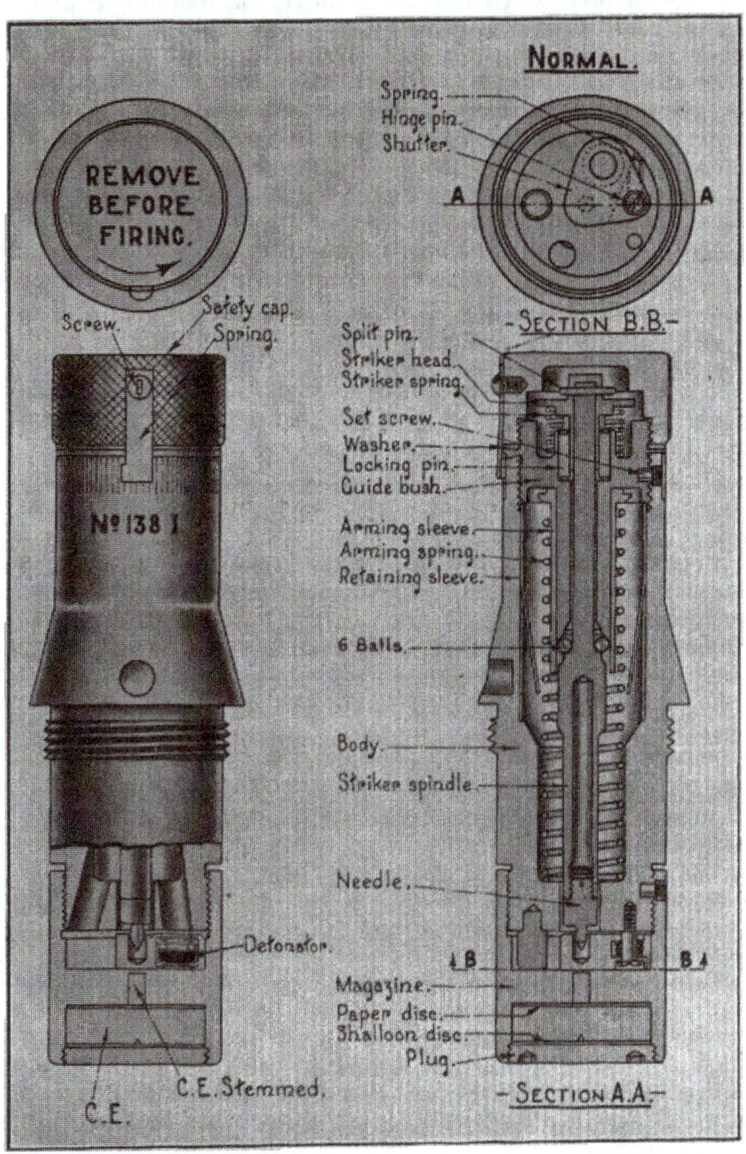

FIG. 15

It is bored from the bottom, leaving a diaphragm between the upper and lower recesses. The diaphragm is drilled almost through the centre from the underside, leaving a thin partition. This channel is filled with C.E. stemmed in, whilst the chamber below it contains a C.E. pellet. The bottom of the magazine is closed by a plug.

The *shutter* is pivoted to the body above the diaphragm and, until the fuze has armed, is held in the closed position by the point of the striker needle engaging in a recess in the shutter. The shutter carries a detonator and is provided with a spring to bring it to the firing position when released by the forward movement of the needle.

The *detonator* consists of a copper shell containing five grains of fulminate of mercury on the top of which is placed a brass disc and cupro-nickel washer, secured in position by turning over lugs formed on the lip of the shell.

The *guide bush* is in the form of a head with shank. The head is prepared with a flange which bears on the top of the body when the guide bush is screwed into it. The head is screw-threaded above and below the flange, the former to receive the safety cap and the latter to screw into the body.

It is bored through the centre for the striker spindle, while the top and bottom of the head portion is recessed to form seatings for the striker spring and arming spring respectively.

The *striker* is assembled with its steel spring under compression between the underside of the striker head and the top recess in the guide bush. The striker head is attached to the striker spindle by a split pin and the top surface is shaped to form a key which corresponds with a key-way in the safety cap. The underside is extended downwards to form a sleeve which fits over the striker spindle and is connected to the guide bush by two locking pins fitting into semi-circular grooves prepared in the sleeve and guide bush. The locking pins allow vertical but not rotary movement of the striker.

The striker spindle is enlarged near its centre, the enlargement being sloped to two diameters at its upper end to form a seating for the balls. The lower end of the spindle is bored and threaded to receive a steel needle.

The *arming sleeve* is cylindrical in shape with a flange formed at its upper end to act as a bearing for the top end of the arming spring and to engage under the springs of the retaining sleeve when the fuze is armed.

The *arming spring*, of steel, is spiral in form and is assembled around the arming sleeve and striker spindle, being held between the flange of the arming sleeve and the bottom of the recess in the fuze body.

The *balls*, of steel, are six in number and are held in position, before the fuze is armed, in the seating formed on the striker spindle by the lower end of the guide bush and prevent the needle from being withdrawn from the recess in the shutter until the fuze is armed.

The *retaining sleeve* is cylindrical in shape and is formed with four flat springs on its inner surface which, when the fuze is armed, engage the flange of the arming sleeve.

The *safety cap* has a steel flat spring, with a projection at its lower end to engage the milling on the fuze body, secured by a screw into a slot on the side of the cap. It is bored and threaded internally to screw on to the fuze body. A key-way in the recess engages the key formed on the striker head in the event of the fuze being accidentally armed. Any attempt to unscrew the cap in these circumstances causes the cap, striker head and striker to be locked to the fuze body through the medium of the locking pins and guide bush.

The top of the cap is stamped REMOVE BEFORE FIRING and has an arrow indicating the direction in which the cap should be turned when unscrewing. All stamping is filled in with white. A dermatine washer is fitted below the safety cap.

Action (Fig. 16)

Before loading, the safety cap is removed.

On firing, the acceleration of the bomb in the bore causes the arming sleeve to set back until the flange on the sleeve engages under the springs of the retaining sleeve, thus compressing the arming spring and freeing the balls from their seating between the striker spindle and guide bush. The striker spring, bearing against the underside of the striker head, carries the striker forward until the seating vacated by the balls bears against the lower end of the guide bush, withdrawing the needle from the recess in the shutter. The latter is thus allowed to be brought to the firing position by the shutter spring and the detonator is then in line with the central channel in the diaphragm of the magazine.

During flight, the striker is prevented from setting back by its spring.

On impact, the striker is forced in, compressing the striker spring and causing the needle to pierce the detonator. The resulting detonation passes to the stemmed C.E. in the channel and to the C.E. pellet in the magazine, from thence to the exploder in the bomb which, in turn, detonates the bursting charge.

FUZE, PERCUSSION, D.A., No. 150, MARK IZ

The *fuze* (Fig. 17) consists of the following principal parts :—

Body, striker guide and set screw, striker with spring and retaining pin, shutter with shutter spring, detonator, plunger with spring, sleeve, cover plate, detent with spring and plug, bottom cap and safety cap.

The *body*, of aluminium alloy, is formed with a head the lower portion of which is screw-threaded to receive the striker guide. The upper portion is plain and is bored transversely to receive the plunger and spring. Below the head it is screw-threaded for a portion of its length to the 1·375-inch fuze-hole gauge. The lower end is reduced in diameter and screw-threaded to receive the bottom cap.

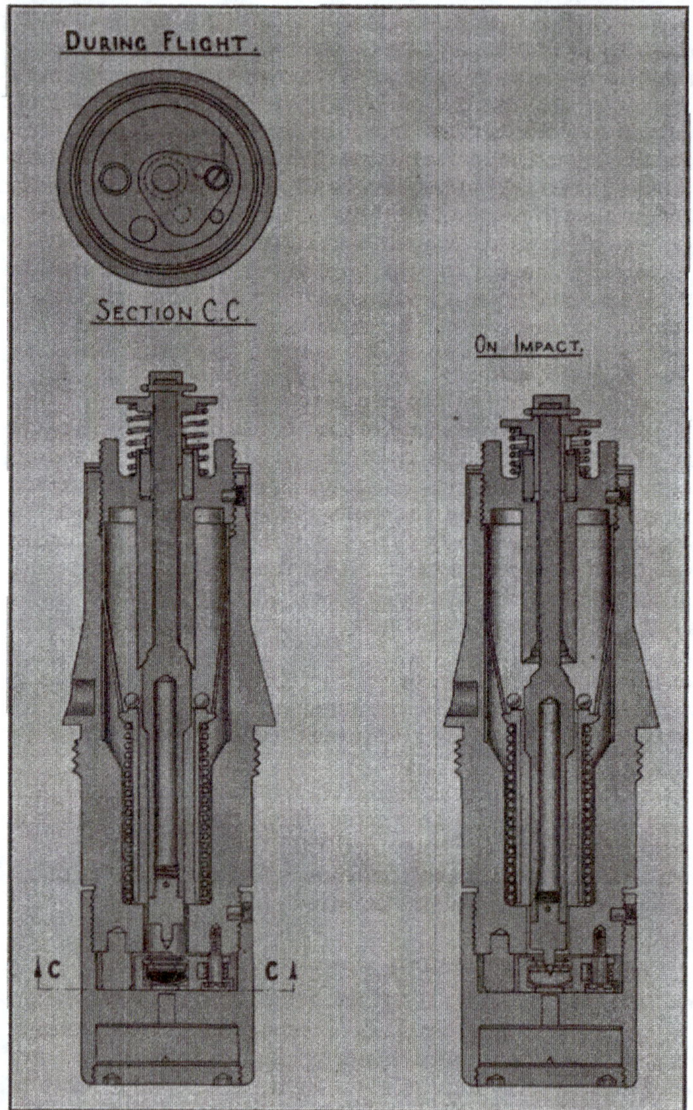

Fig. 16

The top of the body is formed with a groove to accommodate the shutter and a hole on either side of the groove is fitted with a locating pin to retain the cover plate in position.

Internally, the body is bored out from the bottom to form a magazine and a channel is bored centrally nearly through the head, a diaphragm being formed between the channel and shutter recess in the striker guide. The channel is filled with C.E. stemmed in and retained in position by a paper disc shellacked on. The

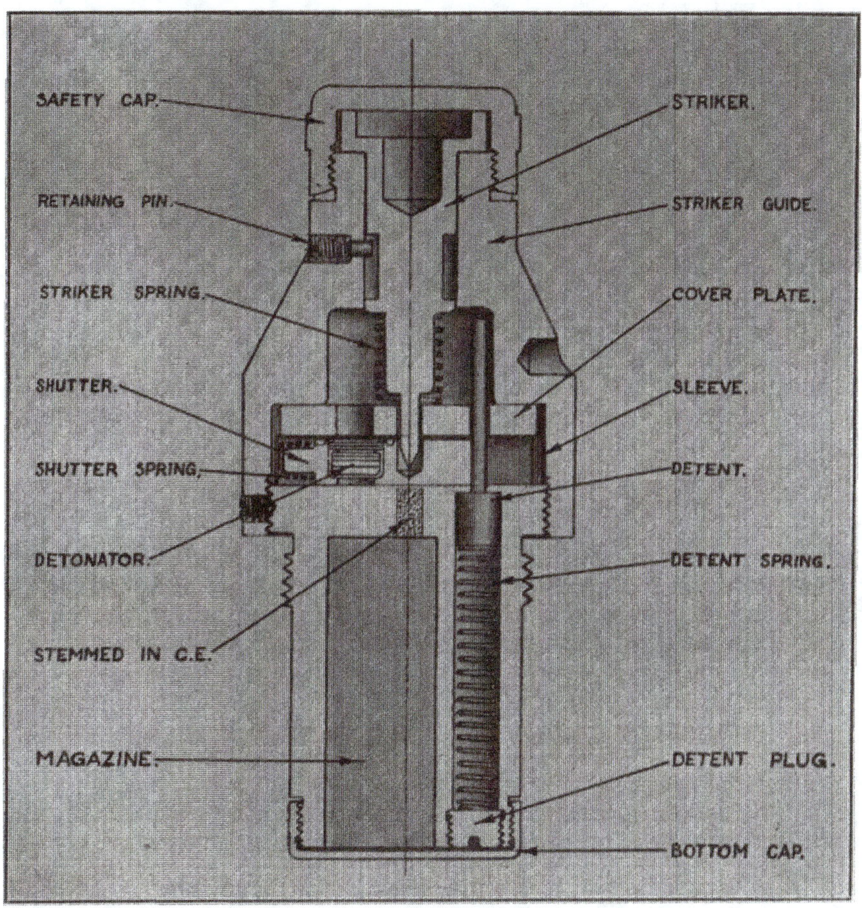

FIG. 17

magazine contains a C.E. pellet retained in position by the bottom cap.

A second recess in the body, at the side of the magazine recess, accommodates the detent and spring, the upper end of the recess is reduced in diameter to form a shoulder and the lower end is screw-threaded to receive a plug.

The *striker guide*, of metal, is dome-shaped and screw-threaded at its upper end to receive the safety cap.   Internally, it is bored out to form three recesses of different dimensions, the lower and largest recess being screw-threaded to screw on to the head of the body.   It also forms a chamber for shutter and cover plate.   The centre recess accommodates the lower end of the striker with its spring, also the upper end of the detent.   The top recess receives the striker body.

A hole is bored transversely opposite the upper recess and is screw-threaded to receive the striker retaining pin and a second hole is bored at the lower end of the striker guide.   It is screw-threaded to receive a set screw for securing the striker guide to the body.

The *striker*, of steel, is formed with a case hardened point at its lower end, above which it is enlarged to three dimensions, the largest dimension forming a head by which the striker is supported on top of the striker guide.   A groove, cut around the body of the striker, is engaged by the inner end of the retaining pin in the striker guide, the pin limiting the movement of the striker when the safety cap is removed.   The striker is supported at its lower end by a steel spiral spring which is kept under tension between the top of the cover plate and the underside of a flange on the striker by the safety cap.

The *shutter*, of metal, is contained in the groove in the top of the body and is formed with a spigot at one end for its spring, the opposite end being concave to form a bearing for the detent.   It is bored and recessed from the top to receive a 5-grain detonator and is recessed to receive the point of the striker which, together with the detent bearing against the end of the shutter, prevents any movement of the latter in its groove until the safety cap is removed and the detent withdrawn within its recess in the body.

The *shutter spring*, of steel, is positioned over the spigot on the end of the shutter and is held in compression between the end of the shutter and a sleeve by the detent bearing against the opposite end of the shutter.

The *detonator* consists of a copper alloy shell containing 3 gr. of lead azide on top of which is pressed 2 gr. of detonating composition (" A " mixture), the whole being retained in the shell by a brass disc and brass washer secured in position by turning over lugs formed on the top of the shell.

The *plunger*, of metal, with its spring, is contained in the recess in the head of the body and bears against the side of the shutter, the spring being held in compression between the plunger and end of the recess in the body.

The *sleeve*, of tinned plate, surrounds the plain portion of the head of the body, closing the ends of the groove containing the shutter.   A bearing is formed in the sleeve for the outer end of the shutter spring.

The *cover plate*, of metal, is secured on top of the body by the locating pins engaging holes in the cover plate. Three holes are bored through on the centre line, one hole being over the detonator, the hole on the opposite side allowing the upper end of the detent to pass through into the recess in the striker guide and the centre hole allowing the lower end of the striker to pass through to its recess in the shutter.

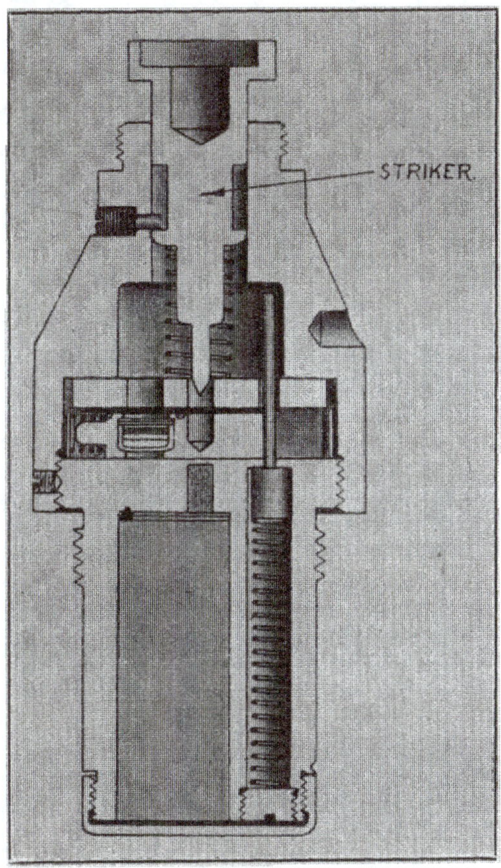

Fig. 18

The *detent* consists of a metal socket with steel stem and torsional spring. The socket is bored and recessed to receive the torsional spring and lower end of the stem, the latter being bulb-shaped at its lower end and bored to receive the end of the torsional spring. The detent is contained in the recess prepared for it in the body of the fuze, the stem projecting through the cover plate into the centre recess in the striker guide and preventing the shutter from moving and bringing the detonator over the channel of C.E. until shock of discharge.

The *detent spring*, of steel, retains the detent in the forward position in its recess until compressed by the shock of discharge. The spring is retained in the recess in the body by a metal plug screwed into the bottom of the recess.

The *bottom cap*, of aluminium alloy, is screw-threaded internally to screw on to the lower end of the body and closes the magazine.

The *safety cap*, of steel, is screw-threaded internally and screws on to the top of the striker guide, protecting the head of the striker. It is milled around its periphery to form a grip when being removed and is stamped on top with the words REMOVE BEFORE FIRING and an arrow indicating the direction in which the cap should be unscrewed, the stamping being filled in with white paint.

A leather washer is interposed between the bottom of the safety cap and a flange formed on the striker guide.

Action

Before loading, the safety cap must be removed.

When the cap is removed (see Fig. 18) the striker spring forces the striker back, releasing its point from the recess in the shutter and hole in the cover plate. The striker is prevented from complete removal from the fuze by the retaining pin engaging the groove around the striker.

On firing (Fig. 19) the shock of discharge causes the detent to set back, compressing its spring and withdrawing the stem from the end of the shutter into the recess in the body. The torsional spring at the lower end of the stem asserts itself and causes the stem to incline over until the upper end is engaged under the shoulder formed in the recess, thus preventing the detent from moving forward and re-engaging the end of the shutter.

The shutter is now free to move in its groove and, under the compulsion of its spring, brings the detonator immediately under the striker point and over the channel of stemmed in C.E.

The plunger, released by the movement of the shutter, moves forward under the compulsion of its spring and engages in front of the shutter, preventing the latter from further movement.

On impact (Fig. 20) the striker is forced in compressing its spring, the striker point piercing the detonator. The resulting detonation blows in the diaphragm in the body of the fuze, passes to the stemmed in C.E. in the channel and to the C.E. pellet in the magazine, blowing out the bottom cap. From thence the detonation passes to the exploder in the bomb which, in turn, detonates the bursting charge.

PLUNGER SPRING
PLUNGER.
SHUTTER.
SLEEVE.
SHUTTER.
DETONATOR.
DETENT.
TORSIONAL SPRING.
DETENT SPRING.

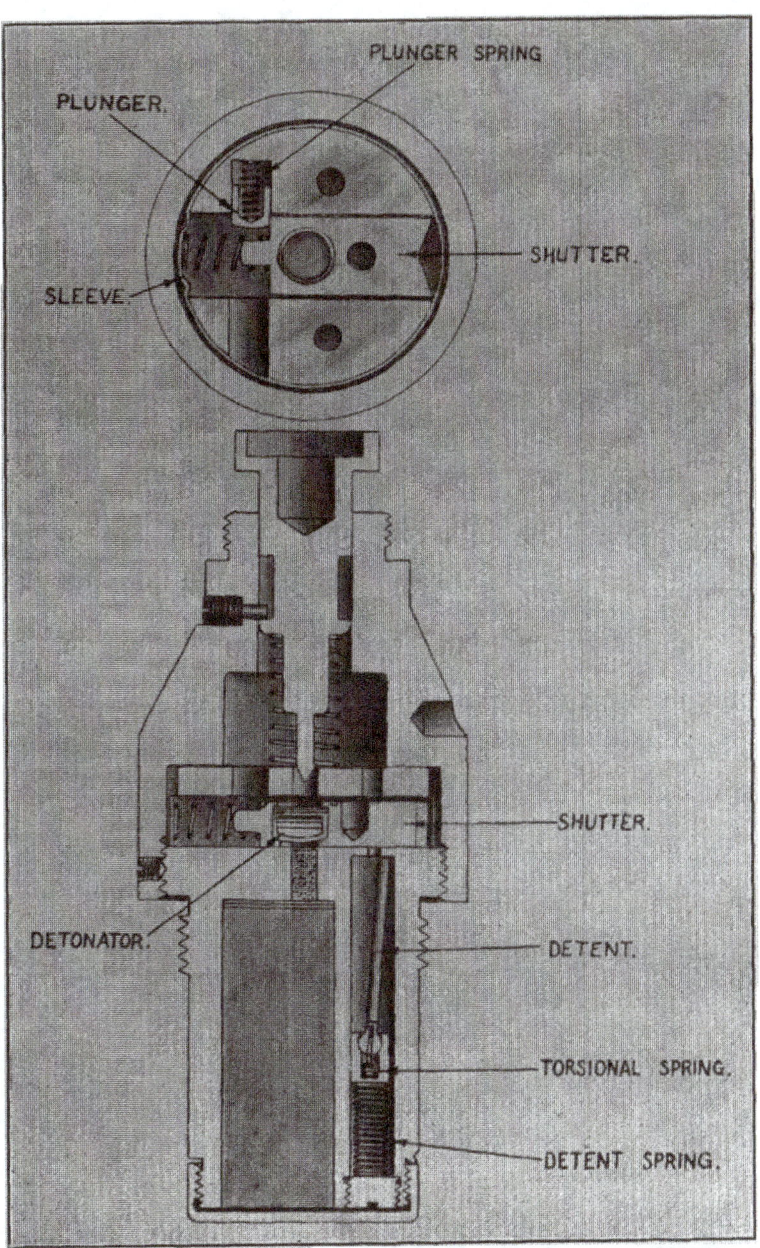

FIG. 19

FUZE, PERCUSSION, D.A., No. 150P, MARK IZ

The *No. 150P fuze* differs from the *No. 150 fuze* in the magazine being filled with about 110 grains of G.12 or R.F.G.2 gunpowder and the channel with G.20 or R.P. gunpowder lightly pressed in.

The fuze is for use in bombs containing gunpowder bursting charges.

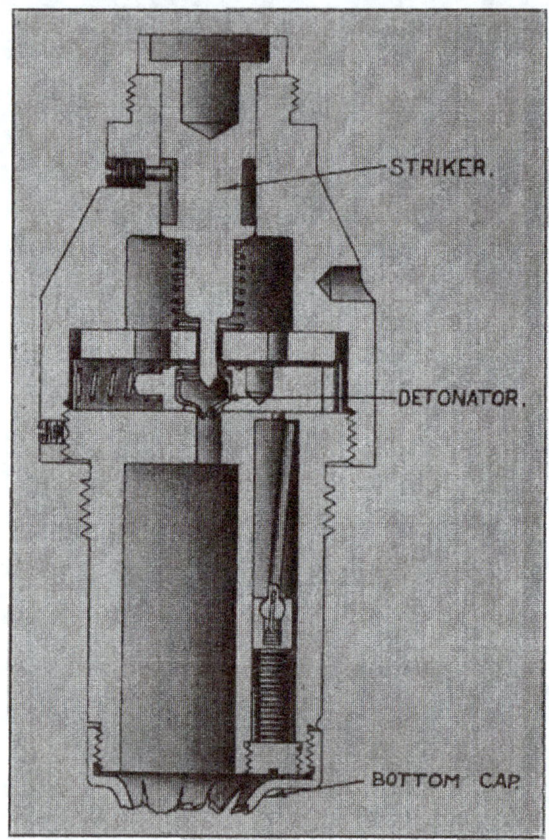

FIG. 20

DRILL AMMUNITION

The greatest care is necessary in the storage and use of ammunition held on charge as drill, every possibility of its being mixed with service ammunition being avoided, as errors of this kind may lead to serious accidents.

Cartridge, Drill, M.L. 3-inch Mortar, 95-Grains, with Striker Clip, Mark I

The *drill cartridge* consists of a yellow paper case and is marked DRILL in red in two places diametrically opposite each other. A fired or empty cap is fitted in the cap chamber and the cartridge is fitted with a wood or rolled paper cylinder internally.

Cartridge, Drill, M.L. 3-inch Mortar, Augmenting, Mark I

The *drill cartridge* consists of a calico bag filled with sawdust and choked at the mouth with sewing silk. It is marked DRILL in red.

Bomb, M.L., Drill, 3-inch Mortar, No. 1, Mark I

The *drill bomb* consists of a cast iron body, steel tail unit and metal nose plug with cap representing the No. 138 fuze.

It is similar in external appearance to the service bomb except that the word DRILL is cast in the body and stamped on the cap, the latter being filled in with white. DRILL is also stencilled on the body in white. The body is painted black.

CHAPTER V

MARKINGS ON AMMUNITION AND PACKAGES

GENERAL REMARKS

Ammunition issued to the service is suitably marked to facilitate identification and to ensure correct segregation in store and transport. Markings also ensure that the correct types are used in the gun and assist greatly in tracing defects in design or manufacture to their source.

Markings are of two kinds, permanent and temporary. The former are in the nature of stamping in the metal of the article concerned and relate, as a general rule, to the process of manufacture, while the latter are usually painted or stencilled and refer more particularly to the explosive elements employed in the ammunition.

Very great care should be exercised in the handling of ammunition, either loose or in packages, to avoid causing damage to the markings, as partial obliteration may render it difficult, and in some cases impossible, to identify the article.

In drawing up a scheme of marking, the general principle observed is that it should always be possible to identify, from the information on the exterior, the contents of a package or a separate article and the conditions of manufacture.

M.L. 3-inch mortar ammunition is batched for the purpose of recording the various components employed in the make-up of the round. Each batch contains a fuze of one lot only, but may contain more than one lot of other components, i.e. the fuzes employed in the make-up of a particular batch will be of one Lot No., but the primary cartridge may be of more than one Lot No.

Batches will be distinguished by consecutive numbers preceded by the appropriate letter, e.g. " Batch B.1 " denotes the first batch of H.E. ammunition.

Batches will be stored separately and will be so arranged as to avoid dividing a batch.

A label is affixed to the inside of each box giving particulars of the components contained in the ammunition therein. The particulars on these labels must be altered as necessary when any change is made in components.

When it is necessary to replace original components by those of other Lot Nos., the letter " X " will be appended to the batch number on the box. It will also be stencilled on the side of the cartridge case when the fuze is changed. The letter " X " denotes that the box contains components other than those originally packed.

In the event of batched ammunition being removed from its boxes and requiring to be re-packed, e.g. unexpended ammunition after firing has taken place, care must be taken to ensure that only ammunition of the same batch is placed in each box.

If possible, the ammunition will be re-packed in the boxes from which it was removed. Failing this, the batch number on the box in which the ammunition is re-packed must be amended to agree with the contents.

The following pages show the markings to be found on M.L. 3-inch mortar ammunition and packages, with their significance.

CARTRIDGES

Stamping on base of primary cartridge
> (*a*) Contractor's initials or recognized trade mark.
> (*b*) *Mark* of cartridge.

Stencilling on top of primary cartridge
> (*a*) Date of filling (day, month and year) in black.
> (*b*) Lot No. of filled cartridge.

Stencilling on augmenting cartridge
> (*a*) *Mark* of cartridge.
> (*b*) Contractor's initials or recognized trade mark of manufacture of the empty bag.
> (*c*) Calibre, e.g. 3″ MOR.
> (*d*) Weight and nature of propellant.
> (*e*) Initials of filling firm or monogram of filling station.
> (*f*) Date of filling (month and year).
> (*g*) Lot No. of propellant.

NOTE.—The *Mark I* 100-grain cartridge has an adhesive label attached to the top of the celluloid container giving the calibre, weight and nature of propellant, Lot No., filling firm and date of filling instead of the above-mentioned stencilling.

BOMBS

(Plate 4)

Stamping on body
> (*a*) Calibre of mortar.
> (*b*) Weight of projectile.
> (*c*) *Mark* of projectile.
> (*d*) C.S., indicating cast steel, F.S. if forged steel.
> (*e*) Contractor's initials or recognized trade mark.
> (*f*) Date of manufacture of empty bomb.
> (*g*) SMK for smoke bombs.
> (*h*) P for practice bombs.

Stencilling on body in white (except where otherwise stated)

(*a*) Calibre of mortar.
(*b*) *Mark* of projectile.
(*c*) Design No. of method of filling.
(*d*) Particulars of filling (in red) on practice bomb.
(*e*) Monogram of firm or filling station.
(*f*) Date of filling.
(*g*) Series No. of filled lot.
(*h*) PHOS on smoke bomb.
(*i*) Monogram of firm or charging station when filled with phosphorus.
(*j*) Date of charging smoke bomb.

Distinctive markings

(*i*) Heads of bombs.
    (*a*) Painted green denote phosphorus filled bombs.
    (*b*) Painted black denote practice bombs.
(*ii*) Rings around head.
    (*a*) Red ring denotes that bomb is filled.
(*iii*) Bands round head.
    (*a*) Green band denotes amatol filled bomb.
    (*b*) Yellow band denotes practice bomb.

NOTE.—Bombs are coated externally with varnish.

FUZES

Stamping

(*a*) No. and *Mark* of fuze.
(*b*) Initials or recognized trade mark of manufacturer of empty fuze.
(*c*) Date of manufacture (month and year).
(*d*) Initials or monogram of filling factory or station.
(*e*) Date of filling (month and year).
(*f*) Lot No. of filled fuze.

AMMUNITION PACKAGES

M.L. 3-inch mortar ammunition is issued in boxes to hold six rounds complete, or, if for use overseas, six bombs, fuzed or plugged, without cartridges.

The particulars of these boxes are shown in the following table.

No. of box	Contents	Dimensions	Material
B.160	6 rds. each in cylinder No. 278 ...	20·5 in. × 10·7 in. × 14·0 in.	Wood
B.161	6 bombs with shuttered fuzes or 6 bombs plugged ...	20·5 in. × 10·5 in. × 14·25 in.	,,
B.162	6 rds. in two No. 1 carriers	21·1 in. × 10·75 in. × 14·8 in.	,,

When bombs are packed separately in **B.161** box, the cartridges and retaining springs are packed in metal-lined cases.

When bombs are packed plugged, the fuzes are packed in the box detailed below.

No. of box	Contents	Dimensions	Material
F.138	25 No. 138 fuzes in cylinders No. 138F	19·8 in. × 11·125 in. × 8·45 in.	Wood

Markings, usually in the form of stencilling, are placed on the sides and ends of the boxes where they are least likely to receive damage and where they can be seen when stacked. In some cases markings are placed on the lid of the box, but this practice is avoided as far as possible on account of the liability of obliteration by workmen walking on them.

The following markings are to be found :—

Boxes containing un-batched ammunition

(1) *Painting*

    (*a*) Vandyke brown (stained) boxes. All M.L. 3-inch mortar ammunition except as enumerated below.

    (*b*) Green boxes indicate smoke bombs.

(2) *Stencilling*

    On both ends

    (*a*) Fuze No. and *Mark* when bombs are fuzed.

    (*b*) Initials of fuze filling factory or station.

    (*c*) Fraction denoting composition of amatol, e.g. 80/20 when box contains H.E. bombs.

    (*d*) SMK PHOS when box contains smoke bombs.

    (*e*) Number of bombs packed.

    (*f*) PLGD when bombs are packed plugged.

(3) *On both sides*

    (*a*) Design No. of method of filling bomb.

    (*b*) Contractor's initials or monogram of filling or charging station.

    (*c*) Date of filling or charging (day, month and year).

    (*d*) Fuze No. when bombs are fuzed.

    (*e*) Number and nature of bombs packed.

    (*f*) PLUGGED when bombs are packed plugged.

(4) *On top*

    (*a*) O.D., denoting box has been oil dressed.

    (*b*) PHOS and DECK CARGO when box contains smoke bombs.

Boxes containing batched ammunition

The batch letters used on 3-inch mortar ammunition boxes, and their significance, are as follows :—

B	...	...	...	...	... High explosive.
C	...	...	...	...	... Smoke.
S	...	...	...	...	... Practice.

(1) *Painting*

The painting is the same as for un-batched ammunition.

(2) *On both ends*

    (a) Fuze No. as applicable.

    (b) Fraction denoting composition of amatol, e.g. 80/20 when box contains H.E. bombs.

    (c) SMK PHOS when box contains smoke bombs.

    (d) PRAC when box contains practice bombs.

    (e) Batch letter and number as applicable.

    (f) PLUGGED FOR — FUZE when bombs are packed plugged.

(3) *On both sides*

    (a) Batch letter and number as applicable.

    (b) Yellow horizontal line across centre of box.

    (c) Nature of bomb packed.

    (d) Fuze No. as applicable.

    (e) PLUGGED FOR — FUZE when bombs are packed plugged.

    (f) 6 BOMBS IN 2 CARRIERS when bombs are contained in carriers.

(4) *On top*

    (a) O.D., denoting box has been oil dressed.

    (b) PHOS and DECK CARGO when box contains smoke bombs.

    (c) 6 CYLS. NO. 278 when bombs are contained in cylinders.

PRACTICE AMMUNITION

A 2-inch yellow band around centre of box.

Markings on fuze boxes

(1) *On both ends*

    (a) FUZES and fuze No. as applicable.

    (b) Lot No. of fuzes.

(2) *On both sides*

    (a) Contractor's initials or monogram of filling station.

    (b) Lot No. of fuzes.

    (c) Date of filling (month and year).

    (d) FUZES and fuze No. and *Mark* as applicable.

    (e) Number of No. 138F cylinders, as applicable, enclosed in box.

(3) *On top*

O.D., denoting that box has been oil dressed.

Markings on metal-lined powder cases containing cartridges

(1) *On two sides*

    (*a*) Manufacturer's initials or recognized trade mark.

    (*b*) Calibre of mortar.

    (*c*) Number, nature and *Mark* of cartridges packed.

    (*d*) Number of retaining springs packed.

(2) *On lid*

    (*a*) Calibre of mortar and nature of cartridges packed.

    (*b*) SPRINGS, indicating that springs are packed.

    (*c*) O.D., denoting that box has been oil dressed.

LABELS

    (*a*) Packer's label attached to underside of lid.

    (*b*) Two station labels affixed over the junction of the lid and box.

    (*c*) Government explosive and classification label on the side of the box.

    (*d*) Label giving particulars of batched ammunition attached to the underside of lid when such ammunition is packed.

CHAPTER VI

LIST OF STORES ALLOWED FOR EACH EQUIPMENT

NOTE—This list is only amended periodically and is not necessarily in agreement with the latest approved scales. It is not to be used as an authority to demand stores.

Articles	No.	Remarks
Axes, pick, 4½-lb.—		
Heads	1	
Helves	1	
Bags, tool, 3-in. mortar	1	
Binoculars, prismatic, No. 2	1	
Boxes, Marks III to VI field clinometer	1(a)(e)	
Boxes, dubbing or mineral jelly, 1-lb.	1 (b)	
Boxes, spare springs, keep pins, washers and insulators, No. 1	1	
Boxes, store, M.L. 3-in. mortar	1	
Cans, lubricating, No. 3	1	
Cans, oil, ½-pint, Mark II	1 (e)	
Carriers, ammunition, M.L. 3-in. mortar, dummy	16	
Cases, mortar clinometer	1 (c)	
Cases, 3-in. mortar sight	1 (e)	
Clinometers, field	1(a)(e)	
Clinometers, mortar	1 (c)	
Collimators, No. 1	1	
Covers, muzzle, 3-in. mortar	1 (e)	
Harness, mortar—		
No. 1	1 (e)	
No. 2	1 (e)	
No. 3	1 (e)	
Implements, ammunition—		
Extractors, cartridge, M.L. 3-in. mortar	1 (c)	
Keys, No. 119	1	
Megaphones, 10-in.	1	
Mountings, 3-in. mortar	1 (e)	
Bracket, supporting, sight—		
Bolt, clamping—		
Springs (spare)	1 (e)	
Gear, elevating—		
Springs, clamping plate (spare)	1 (e)	
Springs, recoil (spare)	1 (c)	
Ordnance, M.L. 3-in. mortar	1 (e)	
Ordnance, M.L. 3-in. mortar, Marks I and II—		
Studs, striker (spare)	1 (c)	
Washers, breech piece (spare)	2 (e)	
Plates, base, 3-in. mortar	1 (e)	

Articles	No.	Remarks
Pliers, side-cutting, 6-in.	1	
Posts, aiming, B.C. and 3-in. mortar	2	
Rods, cleaning, M.L. 3-in. Marks I and II mortars ...	1 (*e*)	
Screwdrivers, adjusting, sight, No. 1	1 (*e*)	
Scrubbers, brass	(*c*) (*d*)	
Shovels, G.S.	1	
Sights, 3-in. mortar	1 (*e*)	
Bracket, collimator—		
Springs, hinge ... ... (spare)	1 (*e*)	
Bracket, range scale—		
Pin, locking—		
Springs ... ... (spare)	1 (*e*)	
Slider, range scale—		
Clicker, deflection—		
Springs ... ... (spare)	1 (*e*)	
Plunger, deflection worm spindle bearing—		
Springs ... ... (spare)	1 (*e*)	
Spanners, adjustable, 11-in.	1 (*c*)	
Wrenches, breech piece, M.L. 3-in. Mark II mortar ...	1 (*e*)	

(*a*) 1 for each battalion.
(*b*) For two mortars.
(*c*) Carried in tool bag.
(*d*) As required.
(*e*) Carried in store box.

CHAPTER VII

CARE AND PRESERVATION

LUBRICATION

The M.L. 3-inch mortar equipment has a number of parts liable to damage by dirt and rust. Cleaning and oiling are, therefore, necessary to preserve the equipment.

Thorough and frequent lubrication of working parts is essential. Whenever and wherever fresh lubricant is applied, the old should first be wiped or scraped off and the parts well worked to distribute the fresh lubricant.

Lubricant will be used as follows :—

Mineral jelly ...	As a preservative generally.
Oil, C.70	Cleaning bright parts and for general purposes of lubrication.
Oil, mineral, burning	Cleaning off clogged oil or mineral jelly.

GENERAL

When cleaning, oil only should be used and no part will be burnished or polished with any gritty substance such as emery paper or bathbrick.

The mortar should be kept clean and free from rust and debris and the bore slightly oiled.

Removable fittings should be frequently taken apart and examined to see that they are sound and in proper working order. They should be treated with care, violence and jerks being avoided and no unnecessary force employed.

Working surfaces should be well lubricated, work easily and be free from burrs. Any burrs must be removed by an artificer.

If a flaw or crack is observed in a mortar, it will be put out of action and arrangements made for it to be examined by the O.M.E.

The whole of the exterior part of the mortar, except the clinometer plane, will be painted.

The clinometer plane must only be cleaned with oil. Any rust should be loosened with a coating of paraffin and then rubbed off with cotton waste. Care should be taken to prevent the plane being damaged or burred. If the mortar is not likely to be used for a considerable period the plane should be coated with mineral jelly.

Before firing, the bore will be thoroughly wiped out to ensure that it is quite clean and dry, and will be visually examined.

At the close of each day's firing the bore will be thoroughly washed out with fresh hot water and allowed to drain. It should then be dried and, when cool, well oiled, the oil being applied by means of a cloth tied over the cleaning rod.

Should the bore be very dirty, it may be cleaned out with paraffin, after which it will be wiped dry and oiled.

Soda in any form must on no account be used for cleaning the bore.

Before washing out the bore, the breech piece with striker stud should be removed. They must be cleaned and oiled, and, when reassembling, it should be noted that the copper washer is in position and the breech tightly screwed home.

The muzzle cover should be kept in position on the muzzle when the mortar is not being used. In wet weather this is of special importance, as water in the bore seriously affects the shooting.

The mounting must be kept clean and all working parts such as the elevating, traversing and cross-levelling gears well lubricated. The recess in the base plate for the rear end of the breech piece should be kept free from earth. Locking bolts must be examined to ensure that they have not worked loose and that the locking pins on the stays are functioning correctly.

SIGHTING

The sight must be handled with great care at all times and frequently inspected to make sure it is free from damage. All parts are to be kept clean and working parts well lubricated with clean oil.

The sight supporting bracket must be absolutely clean and free from burrs.

The sight must not be taken apart unnecessarily. If for any reason it is taken to pieces, it must be carefully treated in replacement and, if necessary, tested.

The sight and clinometer must be kept in perfect condition and adjustment. When not in use they must be kept in their respective cases.

SIGHT TESTS

(1) The sight should be frequently tested to prove that it is in adjustment both for direction and elevation.

    (a) *Errors in direction.*—These may be due to the displacement of the WORM WHEEL BRACKET.

    (b) *Errors in elevation.*—These may be due to the RANGE SCALE BRACKET or RANGE SCALE SLIDER being bent or damaged.

(2) *To test for direction*

    (a) Stores required :—

        (i) Four direction posts (these are required to suspend the plumb lines).

        (ii) Two plumb lines.

(b) Method

Choose an aiming point, preferably a distant one, on to which the mortar can later be layed. Suspend two plumb lines about five yards apart ; and align them accurately on to the aiming point selected.

Mount the mortar at a distance of about ten yards in front of the plumb lines. The base plate must be moved and the barrel traversed until the plumb lines accurately bisect the barrel throughout its length. The barrel is now vertical and layed on the same point as the plumb lines. Now bring the cross-level bubble central, and check the aim with the sight. If the sight is in adjustment, the aim will be on the same point as the plumb lines. If it is not, the sight requires adjusting for direction.

(3) *To adjust for direction*

Loosen the three screws in the elongated slots in the worm wheel bracket, and move the collimator bracket in the required direction until the sight is layed on the aiming point. Having made the adjustment, tighten the screws. The aim should then be finally checked.

NOTE.—The three screws referred to are those of which the heads project above the worm wheel bracket.

(4) *To test for range*

    (a) Stores required—Field clinometer.

    (b) *Method*

      (i) Lay the mortar at the elevation corresponding to 1,000 yards range for full charge. This is done as at (ii) to give an accurate reading for 1,000 yards range and to reduce the inaccuracies of the range scale, at other readings, to reasonable limits.

    (ii) Lay the mortar for elevation at an angle of 69 degrees 47 minutes. To do this, set the clinometer at the corresponding figure and place it on the clinometer plane on the barrel. Position the latter by the operating handle until the clinometer bubble comes central.

    (iii) Bring the longitudinal bubble of the sight central. The sight should now read 1,000 yards on the full charge scale. If it does not give this reading, adjustment is necessary.

(5) *To adjust for range*

Position the range scale reader for full charge. Slacken the wing nut of the range scale slider and bring the longitudinal bubble central. Tighten the wing nut.

Slacken the four screws adjusting the range scale reader and position the reader to read 1,000 yards on the full charge scale.

Tighten up the adjusting screws and check that the bubble is still central.

NOTE.—The screws adjusting the range scale reader are positioned two on each side of the range scale slider and pass through elongated slots in the range scale reader.

CHAPTER VIII

STORAGE AND CARE OF AMMUNITION

GENERAL REMARKS

The design of mortar ammunition follows, generally, along well established lines which have proved suitable for other types of artillery ammunition, and it may be accepted as perfectly safe in store, transport and use, provided certain simple and reasonable precautions are observed. No ammunition component can be approved for the Army until its perfect safety and efficiency for ordinary service use in peace and war has been definitely established by careful experiment and trial. As a final precaution, all service ammunition is subject to a rigid technical inspection before being accepted.

On the other hand, it must not be forgotten that the function of an explosive is to explode, consequently ammunition of all kinds must be handled with care.

There is a further reason for careful handling that concerns efficiency, for certain explosives are liable to deterioration through damp, or heat, or the direct rays of the sun, any or all of which may render them ineffective when required for use. For the reason, therefore, of efficiency as well as safety, the utmost attention must be paid by all concerned to the instructions laid down herein in regard to the storage and use of mortar ammunition. Further details on this subject are to be found in Magazine Regulations.

The ammunition portion of the Preface to this Handbook (pages 6 to 13) contains a good deal of useful information concerning service explosive substances and their behaviour under all conditions and this information should be carefully read in conjunction with these notes on the storage of mortar ammunition.

STORAGE

Ammunition should be stored in a separate building which should be dry and well ventilated. The dryness of a building depends greatly upon its proper ventilation and, with this object in view, buildings containing ammunition should be ventilated as often as practicable.

To minimize fire risk, buildings containing ammunition should be provided with a lightning conductor and the interior should be kept clean and free from inflammable material such as oily waste or other articles which are liable to spontaneous ignition. Materials taken into the building for cleaning purposes should be removed immediately the work has been completed.

Boxes containing mortar ammunition should be stacked in groups according to their batch letter, number and sub-batch letter. They will be stacked on their bottoms to a height not exceeding 11 feet, the bottom tier being raised from the floor on battens to avoid injury from damp. To allow of free circulation of air and easy access, the stacks must not be in contact with outside walls and a passage-way of at least one foot should be left between stacks. All essential identification marks on the boxes should be visible and, to assist the easy identification of groups, cards, giving particulars of the ammunition contained in the boxes, should be hung on each stack.

High explosive and practice bombs, whether packed alone or with other components, may be stacked in the same building. H.E. bombs are filled with amatol, which is a comparatively inert explosive but is affected by moisture. It must, therefore, be stored under dry conditions.

Smoke bombs contain white phosphorus which is liable to spontaneous ignition, if dry, when exposed to the air and must, therefore, be stored in a separate building, isolated from other buildings containing ammunition.

The liability to spontaneous ignition is accelerated by heat and stocks should be stored under as cool conditions as possible. They should not be exposed to the sun.

When fuzes and cartridges, both primary and augmenting, are not included in the packages containing the bombs with which they are to be used, they should be stored separately, each nature in its own room.

When accommodation is limited, the G.O.C.-in-C. may authorize the ammunition, with the exception of smoke bombs, to be kept in one compartment providing the amount of explosive contained in the ammunition is small.

The amount that can be termed a " small quantity " must depend on circumstances but, as a general guide, a small quantity is held to mean an amount not exceeding 200 lb. of explosives. As an approximate guide, the amount of explosive contained in 3-inch mortar ammunition is detailed below :—

100 H.E. bombs	116 lb.
100 practice bombs	50 lb.
100 fuzes	$1\frac{1}{2}$ lb.
100 primary cartridges	$1\frac{1}{2}$ lb.
600 augmenting cartridges	$8\frac{1}{2}$ lb.

All packages containing ammunition must, on receipt, be examined to ascertain that their sealing labels on the junction of the box and lid are intact. If they are not intact, the contents should be examined to ascertain that they agree with the particulars stencilled on the box and the packer's label affixed to the inside of the lid. Any discrepancies should be reported immediately.

Packages which have their sealing labels intact must not be opened until required for use.

Defective packages must be repaired before being taken into an explosive store. Empty packages must not be kept in a store in which ammunition is stored.

Records of each nature of ammunition held on charge by the unit, showing details as to the amount of ammunition in store and when received, should be hung up in each store.

No person, except he is on duty, must be permitted to enter a building in which explosives are stored. Before entering he must deposit any matches, lighters, tobacco in any form or other smoking materials outside the entrance and must turn down the ends of his trousers or overalls, unless the turn-up is secured by stitching, to ensure that they are free from mud or grit.

CARE AND PRESERVATION

PACKAGES

Rough usage of ammunition packages and their contents may cause missfires, hangfires, blinds or prematures. Further, ammunition boxes so handled cannot be expected to retain their airtightness, or their contents their efficiency. The greatest care must, therefore, be exercised in the handling of all ammunition or explosives, whether boxed or unboxed.

BOMBS

Bombs should be kept in their carriers unless they are required for immediate use. The carriers will, normally, be kept in the ammunition boxes. Bombs must not be allowed to lay on the ground as dirt or grit will damage the bore of the mortar.

When handling bombs, care should be taken to avoid damaging the vanes as such damage may affect the flight of the bomb.

Blind rounds must be treated as being in a dangerous condition. Such rounds must be destroyed on the spot without being handled. The destruction of blinds must only be carried out by qualified personnel.

If phosphorus filled bombs show signs of not being airtight, they should be isolated and kept under water if possible. It should, however, be remembered that, if the released phosphorus is allowed to dry again it is as liable to spontaneous ignition as before. Leakage is indicated by the characteristic smell of phosphorus and the presence of white fumes.

FUZES

When once a fuze is exposed to the atmosphere it begins to deteriorate.

All fuzes in mortar bombs are provided with safety caps for safety in storage, transport and handling. These should only be removed immediately before loading.

If the cap of a No. 138 fuze cannot readily be removed by hand, no undue force is to be exerted and the fuze must not be removed. The complete bomb is to be carefully set aside for examination by an inspecting ordnance officer.

Beyond removing the cap, a fuze must not be interfered with in any way. If the fuze is tampered with it is liable to be rendered dangerous to handle and may result in premature explosion.

When inserting fuzes in bombs that have been issued plugged, care must be taken to see that the threads of both the fuze and the bomb are free from grit or dirt. The threads of the fuze, with the exception of the three threads nearest the head of the fuze, will be coated with thin lead-free luting and the remaining three threads with thick lead-free luting, a fillet of the latter also being placed round the flange in the nose of the bomb.

When fuzing bombs, the operation is carried out by two men. One holds the bomb whilst the other removes the plug and inserts the fuze with the appropriate key. The implement for use when fuzing bombs is :—

Key, No. 119 ...　　...　　... For No. 138 fuze.

CARTRIDGES

After removal from the package, care should be taken in handling the bomb to prevent accidental ignition of the primary cartridge or damage to the augmenting cartridges.

The *Mark I* 100-grain N.C.(Y) augmenting cartridge is contained in a celluloid container which should be treated with great care to avoid damage. If the container is broken, the cartridge should not be used but should be set aside for examination.

ORDNANCE, M.L. 3-INCH MORTAR, MARK II
ON
MOUNTING, 3-INCH MORTAR, MARK I
GENERAL ARRANGEMENT

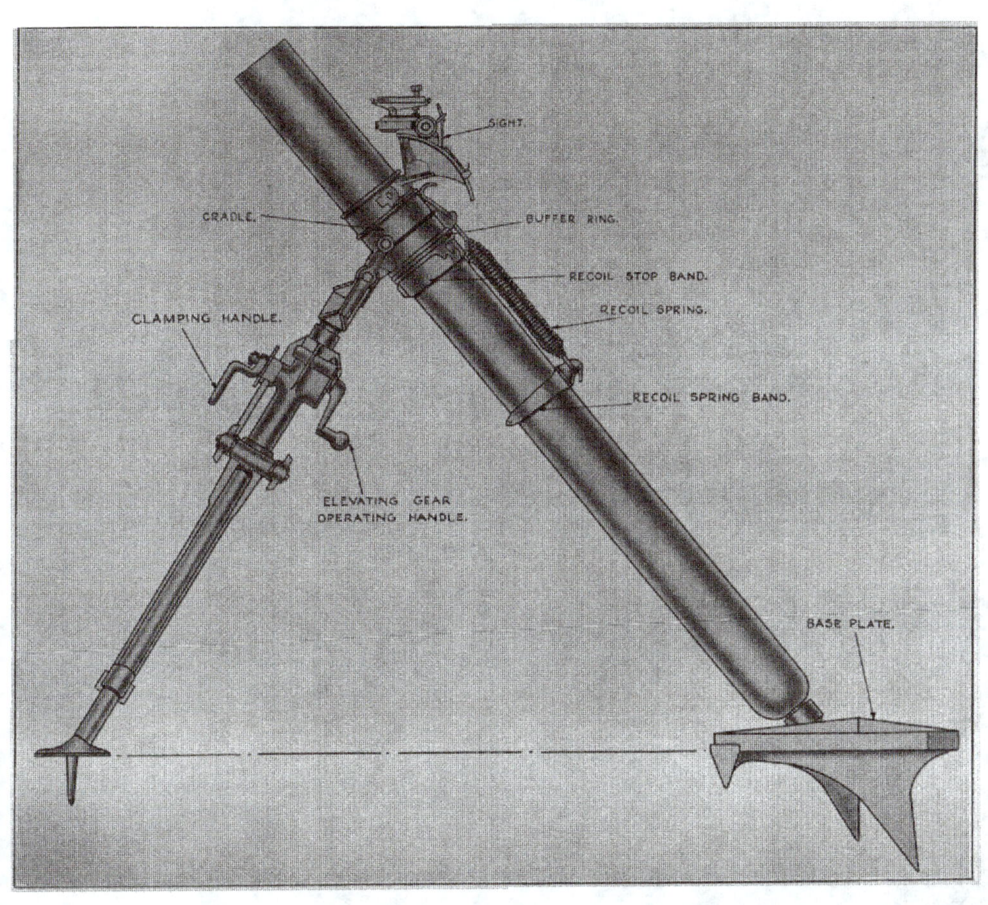

MOUNTING, 3-INCH, MORTAR, MARK I

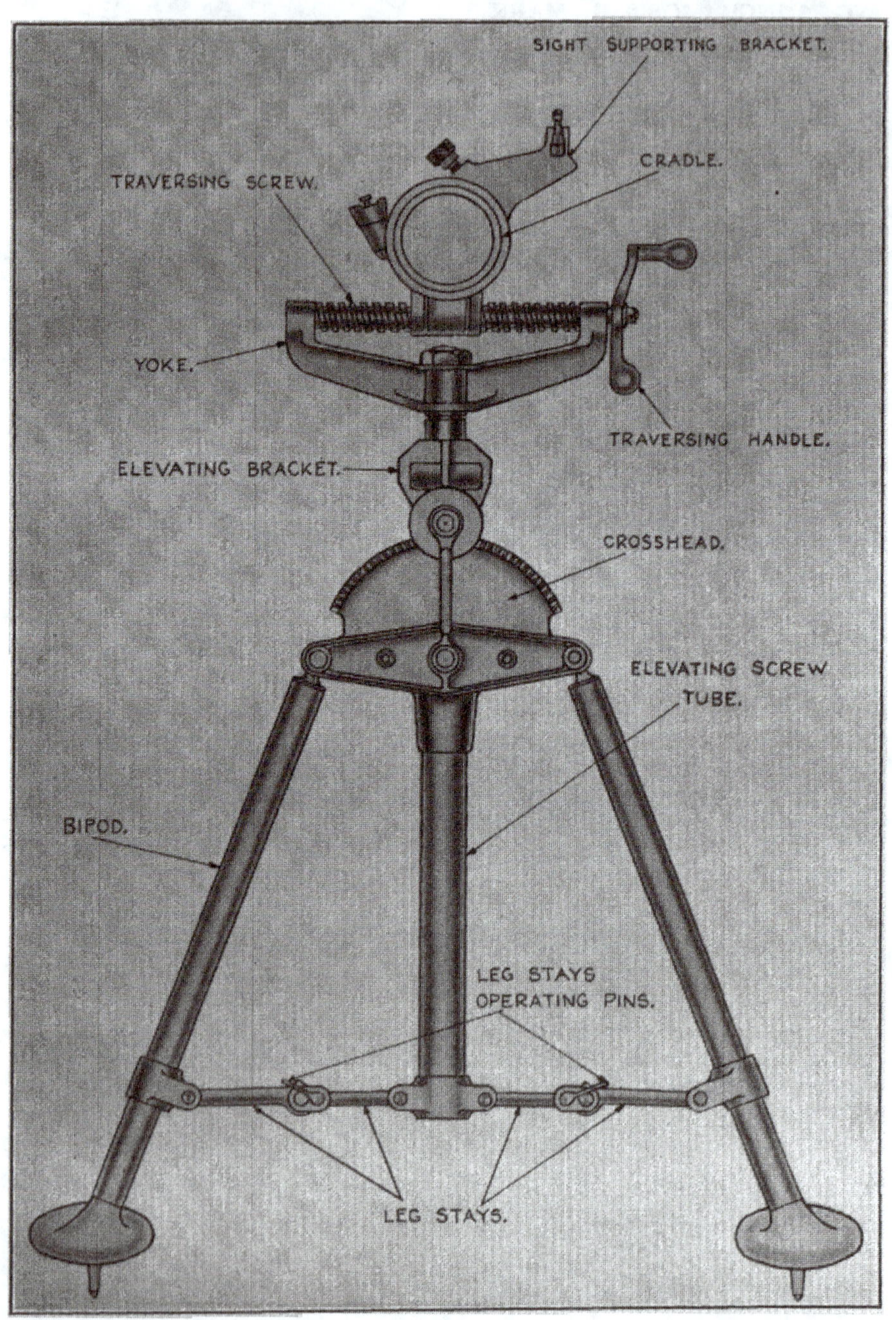

SIGHT, 3-INCH MORTAR, MARK I
ARRANGEMENT

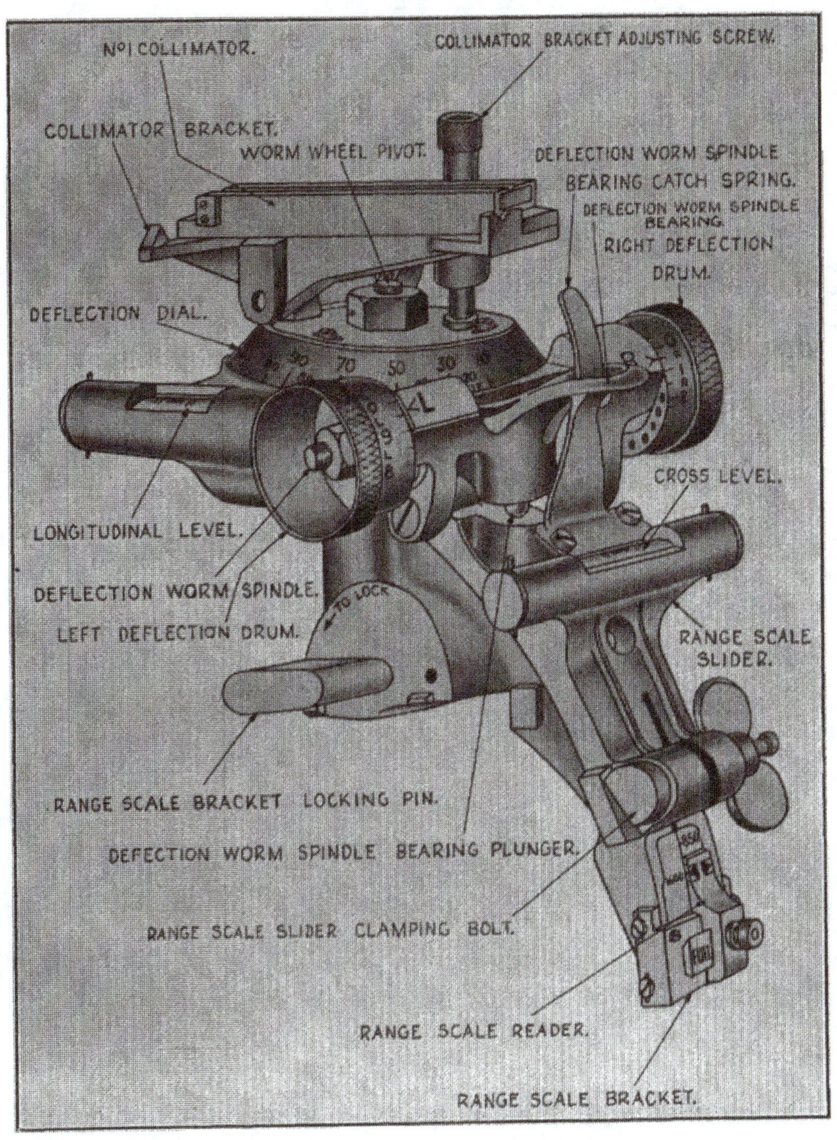

BOMB, M.L. 3-INCH MORTAR.

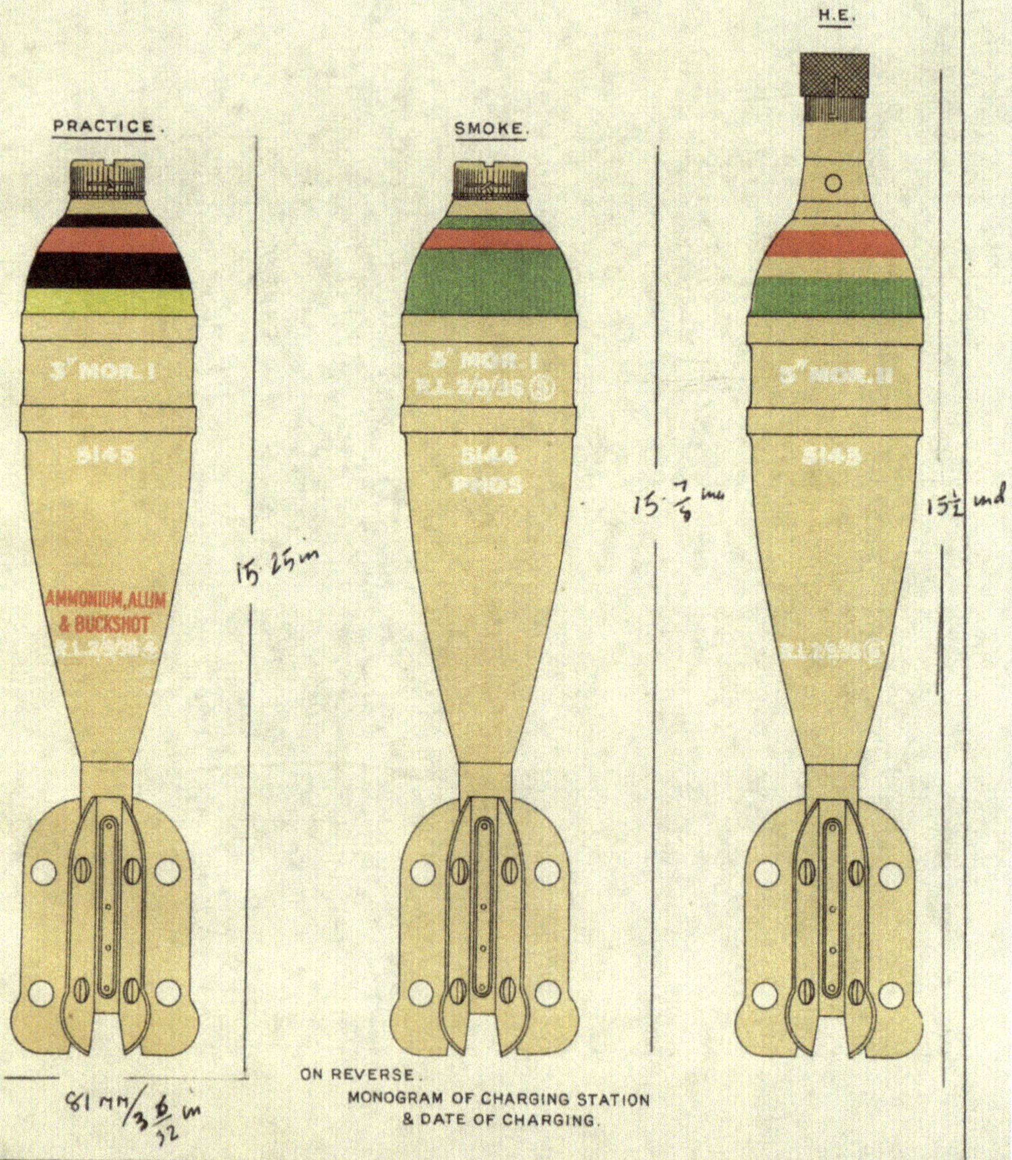

ON REVERSE.

MONOGRAM OF CHARGING STATION
& DATE OF CHARGING.

Malby & Sons, Lith